"[Green's] prose rings with the elemental clarity of the ice he knows so well." —**PEN Awards Committee citation**

"Nature writing of a very high order. . . . A joyride for those who enjoy deep explorations of logic, human frailty and the laws of nature." —*San Francisco Chronicle*

"Brilliant. . . . Resembles at various times the work of Stephen Jay Gould, Loren Eiseley and Barry Lopez, but also Primo Levi's *The Periodic Table* and the poetry of Gerard Manley Hopkins and the writer of Ecclesiastes. It's the kind of book that makes the reviewer want to quote whole paragraphs." —*Plain Dealer*

"Among his many accomplishments, [Green] offhandedly makes the vocabulary of science accessible to the lay reader. He is at ease in the kingdom of poetry—just as much as he is (warily) at ease in the frozen and eerily beautiful Antarctic landscape." —*Boston Globe*

"Some of the prettiest prose ever devoted to the subject of water and lakes and rivers, clouds and rain and fog. A beautiful little book; it will go on the shelf with the other books I read for the love of their words." —*Houston Chronicle*

"A lucid, wondrous account. . . . This authoritative yet lyrical book blends science with art in the enthusiasm that Green feels at being the creator of a new understanding where none was before." —*Winston-Salem Journal*

"Compelling. . . . This book is not only filled with wonder, but also hope." —*Cincinnati Post*

"A magical work of meditation and precise science." —*ISLE: Interdisciplinary Studies in Literature and Environment*

"Poetic and passionate. . . . Green affirms the fact that science, like art, is rooted in pure imagination." —*Booklist* (**starred review**)

water, ice and stone

Science and Memory on the Antarctic Lakes

Bill Green

Bellevue Literary Press
New York

First published in the United States in 2008 by
Bellevue Literary Press
New York

For information, contact:
Bellevue Literary Press
90 Broad Street
Suite 2100
New York NY 10004
www.blpress.org

Library of Congress Cataloging-in-Publication Data

Green, Bill, 1942—
Water, ice and stone : science and memory on the Antarctic lakes / Bill Green. 1st ed.
p. cm.
Includes index.
1. Lakes—Antarctica—Victoria Land. I. Title.
GB1798.V53G74 2008   508.98'9—dc22   2008006367

Bellevue Literary Press would like to thank all its generous donors—individuals
and foundations—for their support.

Book design and type formatting by Bernard Schleifer

Bellevue Literary Press is committed to ecological stewardship in our book production
practices, working to reduce our impact on the natural environment.

♾ This book is printed on acid-free paper.

Manufactured in the United States of America

paperback ISBN: 978-1-934137-08-6
ebook ISBN: 978-1-942658-85-6

3   5   7   9   8   6   4   2

*This book is affectionately dedicated to my wife, Wanda Green,
to my daughters, Dana and Kate,
and to the memory of Eugene and my parents.*

I have been blessed by their company.

contents

acknowledgments

▼

I WOULD NOT HAVE WRITTEN this book without the mountains of encouragement provided by my editor, John Michel. It was he who initiated the project and who saw its potential long before I did. If the book now appears as a coherent whole, this is largely owing to John's steady and gentle guidance and editorial skill. I could not have asked for more.

Valerie Kuscenko at Harmony Books offered many valuable suggestions that materially improved the quality of the book. I am also grateful to her for seeing the manuscript through the many stages of production.

It is a pleasure, also, to acknowledge *The Sciences*, and especially Ed Dobb, for arranging and publishing the journal notes from which this book eventually took form.

Over the years, many friends and colleagues have urged me, through their kind comments, to continue with this avocation of writing: I mention especially Curt Ellison, Harvey Simsohn, Jane Goldflies, Bill Newell, Barbara Whitten, Karl Schilling, Elizabeth Duvert (I thank Elizabeth also for the wonderful quote, once posted on her office door, from Camus. Somehow, I think the book unconsciously took form around those words.), Chris Myers, Hays Cummins, Nancy Nicholson, Muriel Blaisdell, George Stein, and Gene Metcalf.

I am likewise indebted to all the fine people with whom I have had the good fortune to work in the Antarctic. From the early days, these include Roger Hatcher and Robert Benoit; and from more recent times, the members of my own research groups: Don Canfield, Anne Jones, Fred Lee, Mike Angle, Tom Gardner, Tim Ferdelman, Yu Shengsong, Glen Gawarkiewicz, Brian Stage, and Phil Nixon. Other scientists, both at McMurdo and in the Dry Valleys, were extremely generous with their advice and assistance, and, of course, with their companionship and good cheer. George Simmons, Anna Palmisano, Neal Sullivan, Bob Wharton, Berry Lyons, Warwick and Connie Vincent, Art Devries, Irme and Rosalie Friedmann, Diane McKnight, and the late Robie Vestal are among this group. The dedicated and courageous service of the pilots of VXE-6 is deeply appreciated. I want to acknowledge especially the enthusiastic help of the late Steve Duffey.

I am very pleased to thank the National Science Foundation for its support of my research and, indeed, for making travel to the Antarctic possible at all. Frank Williamson, Dick Williams, Guy Guthridge, Polly Penhale, Roger Hanson, Erick Chiang, and Dave Bresnahan have all been extraordinarily supportive.

The scientific encouragement of Gunter Faure, Colin Bull, and Ernest Angino during the early stages of our work on Vanda is deeply appreciated. Keith Chave and Lorenz Magaard arranged a one-year research appointment for me in Oceanography at the University of Hawaii. It was a wonderful year and I am grateful to them for their friendship and hospitality.

I am indebted to Larry Varner for sharing so many of his detailed memories and insights about the Antarctic experience and for carefully reading and commenting on the manuscript during its various incarnations. Jeremy Varner also provided many useful suggestions on the early drafts of the book, as did my friend and colleague Don Canfield. I am pleased to thank three of my former students, Cecile Mariano, William Kennedy, and Jon Reilley, for library research. And I thank Betty Marak for typing and correcting the manuscript.

My sister, Elizabeth Hart, has been a constant source of professional and emotional support, and I wish to acknowledge this here. Dana and Kate, my daughters, have been cheerfully accepting of my long absences and my sustained periods of high distraction. The glazed look of the writer became a familiar one around our house. But somehow they managed to turn all of this to their advantage by realizing that the best time to ask their father's permission for anything ("May I have twenty of my closest friends in for a sleepover?" "May I eat that half-gallon of chocolate ice cream in the freezer?") was when he was writing.

This book could not have been written without the boundless patience, understanding, and critical insight that my wife, Wanda Green, has provided over the years. Her willingness to discuss this project from the very beginning, to read countless rough drafts in all stages of scrawled disorder, and to somehow remain cheerfully optimistic when I, myself, had lost faith that I would ever finish, sustained me throughout this work. To her I am grateful beyond words.

A Lake is the landscape's most beautiful and expressive feature. It is the earth's eye, looking into which the beholder measures the depth of his own nature.

—HENRY DAVID THOREAU

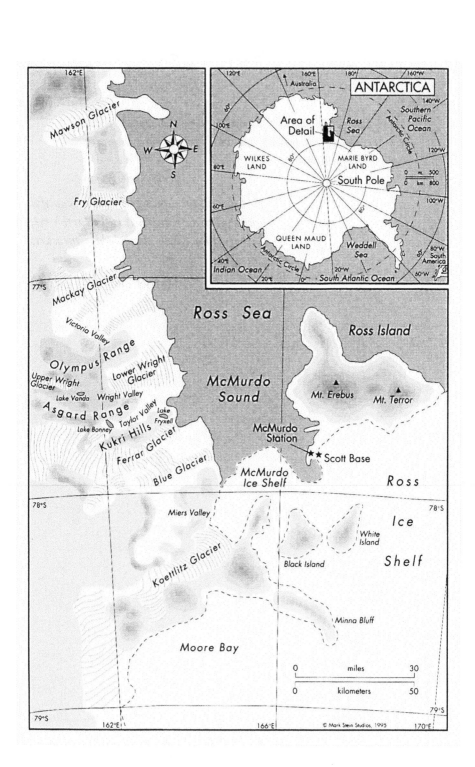

162°E

Mawson Glacier

N
W   E
S

Fry Glacier

120°E   160°E   180°   160°W
Australia
ANTARCTICA
140°W
60°
Southern
Ross   Pacific
Sea   Ocean
Area of
Detail
100°E
Antarctic Circle
WILKES   MARIE BYRD
LAND   80°   LAND
120°W
0   mi.   500
80°E   South Pole
0   km   800
100°W
60°E
QUEEN MAUD   Weddell
LAND   Sea
80°W
South
20°E   Antarctic Circle   America
Indian Ocean   20°W   60°W
20°E   South Atlantic Ocean

77°S

Mackay Glacier

Ross   Sea

Ross Island

Victoria Valley

Olympus Range

Lower Wright
Glacier

Upper Wright
Glacier
Lake Vanda   Wright Valley

McMurdo
Sound

Mt. Erebus   Mt. Terror

Asgard Range
Lake
Taylor Valley   Fryxell
Lake Bonney

Kukri Hills

Ferrar Glacier

McMurdo
Station

Scott Base

Blue Glacier

McMurdo
Ice Shelf

Ross

78°S   78°S

Miers Valley

Ice

White
Island

Shelf

Koettlitz Glacier

Black Island

Minna Bluff

Moore Bay

0   miles   30

0   kilometers   50

79°S   79°S
162°E   166°E   © Mark Stein Studios, 1995   170°E

introduction

An Update for the Age of
Global Warming

Bill Green

WHEN I BEGAN THIS BOOK in the late 1980s, Antarctica, and certainly the McMurdo Dry Valleys, lay on the periphery of modern consciousness. My interest in writing it grew from a personal need to show something of the "human face of science" and to provide an account—all too rare in our literature—of how it felt to do field research in the hostile, austere, but beautiful environment of the Antarctic continent. I had attempted to summarize these motives earlier, in a journal entry for *The Sciences* that read: "Science, sometime in the past, had decided not to deal with certain aspects of reality. The scientist worked with and described the world as abstraction and left the prickly pear of direct experience to the poet and the painter. It was, perhaps, a happy division of labor, but at times it seemed stifling to me. In the midst of a sentence that read, 'Water samples were collected with a 6.2 liter Kemmerer bottle attached to a nylon line,' I wanted to say something about the afternoon shadows on the mountains or the murmuring of a distant stream or the way the wind was sapping my strength. I wanted to say something about the way water tastes on an Antarctic lake after a ten-hour day. These things hovered like ghosts around the edges of scientific prose. They formed the private face of science—the human face." In a way, this statement became the template on which the book was written.

In recent years, in the aftermath of important oceanographic and especially atmospheric studies, Antarctica has emerged as a continent central to our understanding of the Earth as a global system. Its ice caps, and the tiny bubbles within them, have recorded traces of ancient air and have provided clues to past climates. Its stratosphere, and the processes unfolding there, have been probed by balloons and spy-planes and satellites and the data have unequivocally signaled the impact of synthetic chlorofluorocarbons (CFCs) on the Earth's precious ozone shield. The lower atmosphere, above South Pole Station, has provided evidence for the far-flung impact of industrialization and deforestation on the tenuous thread of air—the troposphere—on which all life depends. And the continent's rocks, exposed in only the smallest of deglaciated regions, have been a window onto a deep and warmer past when Antarctica lay in more northerly latitudes.

But, perhaps, most importantly, Antarctica has come to be seen as a key to global climate. We know from studies around the world that mountain glaciers are everywhere receding, whether in the Swiss Alps or on the slopes of Mount Kilimanjaro or in the snowfields outside of Juneau, Alaska, and that sea ice in the Arctic is thinner now and less extensive than ever. In fact, as I write this, satellite images reveal that Arctic sea ice is at its lowest extent ever, raising the prospect of a Northwest Passage entirely open to shipping. Perhaps even more dramatically, we know that the Antarctic Peninsula, which juts northward toward Chile, is one the most rapidly warming regions on Earth. On the peninsula's west coast, mean annual temperatures have increased 2.9 degrees Celsius during the period 1950 to 1999 and winter increases have averaged an extraordinary 5.5 degree Celsius rise over the same period. The retreat of the Larsen-B ice shelf, beginning in 1998, and its rapid collapse in 2002 have been attributed to warming air and water masses in the region. The Larsen-B collapse was vividly described by Christina Hulbe of Portland State University as "a profound event. This ice shelf has endured many climate oscillations over many thousands of years. Now it's gone." Temperature effects on the West Antarctic Ice Sheet and on ice in the vast continental interior are subjects of intense research and debate, and it has been estimated that melting of the ice sheet in West Antarctica alone would contribute an additional twenty feet to sea level—a change that would require the map of the Earth to be redrawn.

The increase in regional and global temperatures—the latter predicted to increase by 1.4 to 5.8 degrees Celsius by the year 2100—is no longer a mystery. Human activities, from the Industrial Revolution to the present, have altered our atmosphere by, in effect, vaporizing long-stored deposits of coal and oil and sending them skyward transformed as the greenhouse gas carbon dioxide. The long Antarctic ice cores, collected near Vostok Station, show that at no time in the 650,000 years prior to the Industrial Revolution had the concentration of this gas been higher than 300 parts per million (ppm). Today, carbon dioxide values stand at 380 ppm, and rising. In addition, our varied activities across the globe—cutting forests, planting rice, developing new products, expanding agriculture—are contributing even more greenhouse gases, like methane, nitrous oxide, and CFCs. As Alan Weisman has noted in *The World Without Us*, "Among the human-crafted artifacts that will last the longest after we're gone is our redesigned atmosphere."

Of all Earth systems, the atmosphere has always been the most vulnerable. The German astronaut Ulf Merbold called it "a fragile seam of dark blue light," and geochemists have long commented on its relatively tiny mass when compared with the oceans and the Earth's crust and mantle. It was in the atmosphere that we first noticed change on a truly global scale, with James Lovelock's observations that chlorofluorocarbons had migrated from the northern to the southern hemisphere and that most of the tonnage released, largely by aerosol spray cans at that time, had remained unchanged in the troposphere. Sherwood Rowland and Mario Molina predicted in the early 1970s that the fate of these compounds would be to rise into the stratosphere and, once there, react with ozone. It took more than a decade for this prediction to be confirmed by Joe Farman, Susan Solomon, and others over the Antarctic continent. A series of conferences, beginning in Montreal, resulted in a ban on chlorofluorcarbon production, but the long lifetime of these compounds (measured in decades to more than a hundred years) ensures that the problem of ozone depletion will be with us well beyond mid-century.

The illusion that human beings are insignificant actors on the global stage, in comparison with the mighty forces of nature, has taken some time to fade. I recall reading an essay by the great explorer Thor Heyerdahl about one of his last ocean voyages. In it, he spoke of how everywhere on the open sea he could discern the signature of man, most disturbingly in the form of the unsightly tar balls that washed against his raft. His point was that what we had once considered to be limitless—the uncharted seas of Cooke and Melville—were nothing more than expansive, closed-basin, lakes, finite in size and corruptible by our collective onshore activities. There has been a shift in our conceptual geography: from thinking of the Earth as being limitless and vast beyond our comprehension to being what it truly is, "a pale blue dot" in space, finite and vulnerable to the prodigious force that we—the six billion of us—have become. The chlorofluorocarbon story amply confirms this, as does the sadly personal and prescient account of climate change that Bill McKibben has offered in *The End of Nature*.

▼ ▼ ▼

Water, Ice and Stone is set in the McMurdo Dry Valleys, a place where it may still be possible to imagine that the Earth abides unchanged. I recall once having climbed to a certain height in the Asgaard Range far above the floor of Wright Valley and the frozen surface of Lake Vanda. When I turned around and sat for a moment to rest, I was struck by the fact that I had been here, in this same spot, nearly thirty years before and that nothing had changed. Nothing. The same river, the same lake, the same blue, flawless sky, the same untouched land, the cold. It was a happy thought. And yet it lasted only a second. For indeed much had changed. The ultraviolet radiation from the sun now poured over me with a far greater intensity than before; and the very air that I breathed had been transformed over those decades so that it now contained some thirty percent more carbon dioxide than it had when I last had this view. While I could not detect the change, all the instruments agreed: things were different now, and I knew that we human beings, a major force on our small planet, were the cause. Though I could not detect it with my senses, the air I breathed was in some way, as McKibben noted, artificial, man-made. Even here!

The McMurdo Dry Valleys are an atypical region of the Antarctic continent. Slightly greater in area than the state of Rhode Island, they seem little more than patches of dry earth pressed between the silver expanse of the Ross Sea and the vast interior ice sheet of East Antarctica. Most of the writings about this place have been the writings of scientists. There has been no diarist to chart this wilderness, no poet to capture its solitudes. Between the early explorers and the few recent essayists, painters, and landscape photographers, the valleys have been described largely in the technical literature of biologists, geochemists, geologists, and paleontologists. These are the people who have walked this land, who have labored and camped in it, and felt it in their bones. What has been communicated to the public has been done through the rare television special or through the photographs of artists like Craig Potten or Eliot Porter.

But the technical summaries of scientific research offer a portrait of the land that is rich and compelling and that sets a broad universal context for what has been learned and carried away from this remote place. For the valleys are treasure troves of information about ancient landscape processes and about the quality of life *in extremis*, where, somehow, living matter gains a foothold in rock and ice and adapts, flourishes even. To know that in seemingly lifeless soils there exists a community of nematodes that live among yeasts and bacteria and filamentous fungi in a frightening matrix of coarse stone and bitter cold is to be reminded of life's tenacity, its elemental toughness.

In the technical scientific literature, a curiously poetic term has been used to capture the essential mix of elements that interacts and gives identity to the McMurdo Dry Valleys and that sets them apart from landscapes anywhere else on Earth. The word is *mosaic*, and it seems an appropriate one, for, in one of its senses, it means a pleasing picture or design made by arranging small bits of colored stone or glass in a matrix. When the term is applied to the Valleys, it refers to the inlaying of glaciers, ephemeral streams, permanently ice-covered lakes, exposed bedrock, sandy soils and richly patterned ground. And perhaps more insubstantially to the vast exposures of geologic time and to the ever-present wind, which works its way into the psyche of anyone who has ever spent a season in the mountains or on the surface of a lake. These are the figurative "bits of stone," the "shards of glass," the raw elements of place and identity. To enter this land is to become part of a mosaic that is at once unexpected and frightening, beautiful and sheltering, timeless and yet touched with the tracings of time. Like William Blake's world in a grain of sand, it is possible from this remote vantage point to envision—in fact, to experience—the turning of the Earth's great geochemical cycles.

▾ ▾ ▾

In the years since the first printing of *Water, Ice and Stone*, the Dry Valleys have assumed a far greater scientific importance. They have become the setting for intensive investigations under the auspices of the National Science Foundation's Long Term Ecological Research Program, and the biologist Edward O. Wilson begins his *Future of Life* with a description of the extremes of aridity and temperatures that challenge even the hardiest microscopic organisms in this setting. "On all of the Earth," Wilson writes, "the McMurdo Dry Valleys most resemble the rubbled plains of Mars." Moreover, the ice-covered lakes, which are pellucid and striking and so oddly out of place in this driest of deserts, serve as indicators of climate change, both in the thickness of their ice and in the structure of their salinity profiles. On timescales of decades to millennia, the lakes are unique ledgers of subtle or pronounced changes in the climate of the region. Understanding what they are telling us is more important now than ever.

While this book was not intended to be a study of climate change, it was meant to convey the many linkages that exist—through the cycles of water, carbon, calcium, oxygen, and many other elements—between ourselves and the Earth. In some ways it became a meditation on this very subject. The chapters "Science and the Shell," "The Sea," "The Flume," The Cone of Erebus," and "Cathedrals" all refer explicitly or implicitly to the great carbon cycle which links us and all of the creation to the solid Earth and to the sea and the atmosphere. It is the carbon cycle that we have so profoundly altered, and these alterations will be the concern of our own and of generations far into the future. We are only now beginning to recognize this.

In the chapter "The Moat," the seasonal shape of the carbon dioxide curve—a subject explored in Al Gore's *An Inconvenient Truth*—is described in global terms:

All over the world, this shallow breath of the Earth is signed upon the air. The signature rises and falls through time, in a sinuous curve. You could not mistake it for anything else but the Earth's breathing. It is the curve of carbon dioxide through modern time, its concentration in the atmosphere, plotted month after month—the Keeling Curve. From the slopes of Mauna Loa on the Big Island to the desolate wyomings of Pole Station— where the sleepless recorders work day and night inhaling the air, taking its measure—the patterns are the same: In the spring of the northern hemisphere, the carbon dioxide of the atmosphere declines as the trees unfold, as they imbibe water into their roots and carbon into their leaves, as the warming seas ripen with phytoplankton off Peru and California and Cape Cod. The sweet yearly inhalation and budding, fecundity and opulence and color everywhere spreading like a blush and holding. All through spring and summer the Keeling Curve falls. In autumn and winter it rises, when in fields and empty lots, in forests and swamps, and in the gray winter of the sea, carbon dioxide is returned to the air.

These are precisely the connections that have been drawn in recent years by those seeking to model the carbon cycle and to predict the influence of our dependence on fossil fuels.

The rising Keeling Curve is the signature of human activity written on air, and the scientific community no longer doubts the role of our industry, broadly construed, as the engine of atmospheric change. This change is certainly in evidence at the Poles and in receding glaciers across the globe and in migrating species on land and at sea. How much ice we are likely to melt and when—in Greenland, in West Antarctica, and on the huge East Antarctic Ice Sheet—is a subject of intense discussion. What actions our increasingly clear scientific knowledge should require is also a matter for debate. Between what we know and how we act falls the shadow, to paraphrase T. S. Eliot. And policy decisions in the wake of scientific evidence, regardless of how persuasive, have always been difficult to formulate. The phase-out of chlorofluorocarbons took some thirty years after the original work of Rowland and Molina; and carbon dioxide reductions, given the role of fossil fuels in every facet of our economy and in the economies of developing countries, will certainly be a far more vexing problem for the nations of the world to solve. Signals from Antarctica and from other frozen shores may well be the prime movers in a resolute global effort to arrest what has been a long assault on the Earth's most fragile reservoir.

▼ ▼ ▼

Reviewers of *Water, Ice and Stone* clearly saw the environmental concerns that it raised and the cycles that it highlighted. But they also saw more. Perhaps more than I had seen myself. Philip Zeigler, for example, in his remarks for the John Burroughs Association, commented on the idea of *transformation* that runs throughout the book:

> Like all lakes, those of the Antarctic are fed by streams which bring to them the detritus of the surrounding land: nutrients, metals, organic and inorganic elements, which are somehow transformed in their passage through the lake. Because of their very isolation, their detachment from the often cataclysmic changes man is continually making in the physical world, they allow us to trace the way in which the elements of the Earth, dissolved from rock by the action of water,

are reconstituted in new forms. They can provide clues about how bodies of water are formed, how they vanish over time, and what happens to their elements. The central process here is a chemical one, a process of transmutation of earth, water, ice, atmosphere in an endless cycle from one form to another.

While the science of Antarctica has assumed an ever greater significance, in light of human-induced changes to the atmosphere and in light of the virtual certainty of a warming planet, the value of the place as a center for contemplation and reflection remains as it was when I wrote the book. Questions about the nature of science and creativity, the choice of one's profession, the connectedness of things in personal and geological time, the coexistence of beauty and death in the same mix, the distinction between solitude and loneliness, the hidden importance of the global cycles, especially those of water and carbon—these are enduring human concerns.

Water, Ice and Stone is a difficult book to classify. It is a travelogue, a scientific quest, a memoir, a hymn to the water molecule, and perhaps, as one reviewer noted, it is "a braided river" where all of these join as one. In one of his essays, John Burroughs said that "it is sympathy, appreciation, emotional experience, which refine and elevate and breathe into exact knowledge the breath of life." In his own writings, he showed how the two—precise scientific knowledge and personal response—could be joined. It is in this tradition—what has come to be called "nature writing"—that the rational and romantic impulses merge. What is impossible in the scientific paper, the full expression and evocation of one's immersion in a particular place and time, are the sentiments that the naturalist essay welcomes: the cold, the sight of clear water, the sense of presentness, of being alive here, now. What is difficult to achieve in poetry and literature—straightforward discussion of process, cause and effect, a deep appreciation of number, law, and the depth of things—the naturalist essay, as practiced since Thoreau and Burroughs, easily assumes into itself.

It was my hope when I wrote that first journal entry, and it is my hope now, many years later, that *Water, Ice and Stone* will be viewed as a work that brings both of these traditions, the scientific and the literary, together. It is also my hope that readers will find in these pages some reason for optimism. For our studies of the lakes of the McMurdo Dry Valleys have shown once again nature's powerful resilience, have shown what is generally known: "that in every lake and ocean, in every parcel of atmosphere, there is a cleansing that tempers the Earth, that drags it back from squalor, that countervails its self-undoing." I should have said, too, that countervails what we, ourselves, have done. If we are truly lucky, these words, in time, may also apply to the profound alterations we have made, and continue to make, to the global carbon cycle and to the thin band of air—that "fragile seam of dark blue light"—into which it is so tightly woven.

Preface

▼

FIRST WENT TO ANTARCTICA in August of 1968 as a member of a research team led by Dr. Robert Benoit of Virginia Polytechnic Institute and State University. Benoit had been awarded a grant by the National Science Foundation to study the microbiology of a strange group of permanently ice-covered lakes located in the rugged coastal valleys (now called the McMurdo Dry Valleys) near the Ross Sea. The lakes had been discovered by the first expedition of the great Antarctic explorer Robert Falcon Scott (1901-4), but only a few scientists had visited them in the intervening years and little had been written about them in the scientific literature. In 1968, I was a graduate student at work on a project in physical chemistry and nearly a year away from completing my doctoral program. When I heard that Benoit and his graduate assistant, Roger Hatcher, needed a chemist, I quickly volunteered, thinking that this would be a wonderful way to learn something about limnology, the science of lakes, and to explore a part of the world that I would most likely never see again.

From the outset, I found the science, and indeed, everything about the Antarctic continent, fascinating. It was the most austere and beautiful land I had ever seen, and I was drawn to it immediately, in ways and for reasons I did not understand. When I was not analyzing the unimaginably clear waters of Lake Vanda or Lake Bonney, two of the ice-covered lakes we had come to study, I was walking and thinking and writing page after page of notes in my journal, trying to describe what I was seeing and feeling in this improbable Eden of ice and stone. My words seemed always to fall short, and seem to do so to this day.

The years passed after that first encounter, but the lakes and the valleys of Antarctica remained vivid memories, a daydream away. Gradually my professional interests turned from the pure laboratory chemistry in which I had been trained to the science of geochemistry. From the courses that I was teaching and from the Midwestern lakes that I was studying, new questions began to arise. I began to think that the answers might lie in the distant waters of the Dry Valleys that I had once visited with Hatcher and Benoit.

In 1980 the National Science Foundation funded my proposal to study the behavior of nutrients (compounds of nitrogen and phosphorus that control the biological productivity of a lake) and heavy metals (elements such as manganese, iron, cobalt, nickel, cadmium, and lead) in Lake Vanda and its inflow, the Onyx River. This turned out to be the beginning of a ten-year investigation that gradually encompassed more lakes and more subtle and vexing questions. How and when had these odd water bodies evolved? Why were they so different in their chemical compositions? How were they sustained biologically in such a harsh environment? How were the chemical elements transported to them, and how were those elements removed? Because they were so biologically uncomplex, might these lakes not serve as beautifully simple models for other bodies of water on the Earth? Perhaps even the oceans? Because they seemed to cleanse themselves of the metals brought into their waters, might they not tell us something about the way the Earth regenerates itself? These were some of the questions that brought me back to Antarctica five more times during the 1980s and, most recently, from October to December of 1994.

Thus the present book draws on more than fourteen months of journal notes collected over seven field seasons. It arose, I think, from the need to talk about the Antarctic work in a more reflective and personal way—in a way that could not easily be accommodated within the pages of professional journals. From the outset, the continent raised questions for me that went beyond the purview of science. These were questions about the ways we experience the world and respond to its physical settings; how we decide, as individuals, to do with our lives what we do with them; the sources of our wonder; the nature of science itself. If the book appears to have a spiritual dimension, that seems only fitting, given the place in which it was written. For Antarctica is the most sublime of continents, a land of light and darkness, of scrawls and traces and hints of eternity.

The language and ideas of chemistry, geochemistry, and limnology figure prominently in these pages. I hope they will not prove distracting to the reader. I believe there is a certain beauty to the sound of the names of the chemical elements and to the names of common field equipment and laboratory objects—the "trim and tackle" of the work.

And of course the ideas themselves, pure creations of the human mind, are things of uncommon beauty. But this is not meant to be a book about those specific sciences. Nor, despite its reminiscences of family and childhood in Pennsylvania, is it meant to be a memoir. Though somehow, through the detours of writing and the strange ways of the heart, science and memory—indeed, science and life—have tangled themselves together in its pages.

In nature, the great geochemical cycles rarely turn exactly back upon themselves. So it seems with life. In the end, we are home but not in the same place. If this book is in some sense a travelogue, the journey it describes is very much a "heraclitean" one. The stream we had hoped to step in is no longer there.

I should note that our work occurs within a very small and atypical region of the Antarctic continent. The McMurdo Dry Valleys—located about seventy miles west of Ross Island and the major U.S. base, McMurdo Station—occupy only a tiny niche (1,500 square miles in a continental area of 5.5 million square miles) at the edge of the great ice sheet that covers most of Antarctica. The valleys were carved by the advance and retreat of glaciers that moved down from the Polar Plateau, but they have been ice-free for some four million years. In these cold, arid, ferociously windswept, and virtually lifeless valleys lie the lakes and the few ephemeral streams that feed them: Lakes Miers, Fryxell, Hoare, and Vanda are the settings for much of what occurs in the book.

Two limnological terms appear early on, and it may be helpful to define them here. In the temperate zone, deeper lakes tend to stratify in summer. The warmer, less dense, and, typically, oxygen-rich waters that lie near the surface are referred to collectively as the *epilimnion*. The cooler, denser, and often stagnant lower waters are called the *hypolimnion*. This usual stratification pattern—warm waters at the surface, cool waters below—is reversed in the lakes of the McMurdo Dry Valleys.

water,
ice and
stone

ONE

Ohio: West and South

Ohio has had its autumn glory. The leaves of the burning bush beyond the sun room had hung like blades of fire for more than two weeks. At Hueston Woods, from the stream-carved valley of Little Four Mile Creek, you could look up walls of late Ordavician shale and limestone into the canopies of orange maples and plum-colored beech, into a crisp expanse of sky that lay as cloudless as a blue canvas behind the brilliant trees. This morning, though, out over the long lake that cuts northwest through the woods, color is only a memory; the world is gray. Still, I am not regretting what has passed. The electron-rich chlorophylls and beta-carotenes; the skeletons of anthocyanin molecules, these will come again, gathering pale light, returning it to us transformed as gold and magenta in the gladdening liturgy of fall. The world moves, cycles through time, offers its gifts afresh.

▼ ▼ ▼

I AM SITTING ON THE WOODEN DOCK that extends over the drab waters of Acton Lake. The lake itself is nothing unusual, though I often wish that it were. It is not as if, fifteen thousand years ago, the fluted, rock-bearing snout of some south-extended glacier had begun to pool and retreat in the wakening Ohio spring, leaving behind a strand of sweet water—the reflecting waters of mastodon herds and cold adapted spruce. Unlike the fingered extensions of Lake Erie and Lake Michigan, or even the kettle lakes of Champaign County to the north, Acton Lake had no such romantic origins. It was created by human hands, the work of convicts impressed not long ago into the service of the state. Nothing exceptional.

And yet it has all the features of the ideal lake. In summer it stratifies. Layers of warm, light water drift on the cool, denser strata below. These upper waters mingle with the atmosphere. Winds from among the corn rows press the surface, sending molecules of oxygen through the tearing lake skin, riding the currents deep into the mixed layer. Streams bring an abundance of the nutrient elements, especially nitrogen and phosphorus from the neighboring farm fields and silicon from the channel rocks. Carbon, the most important element of all, comes to the lake from the air, from carbon dioxide, and from the weathering of ancient carbonates—carbonates layered long ago by shallow seas that once crept over this land, back when the Earth's surface was arranged in unfamiliar ways. There is nourishment aplenty for the drifting phytoplankton and zooplankton and for the microscopic rotifers turning like pinwheels in the lapping shallows, and on up the food chain to the bluegill, to the rugged, omnivorous carp that wants for nothing. Acton Lake, in the parlance of limnology, is eutrophic, well-fed, teeming with life.

In summer, beneath the warm, oxygen-rich surface, lies another lake. Limnologists call it the hypolimnion, but it is really another world. In the depths of the hypolimnion, water meets mud, not sky; the lake is chill and dark. Things sink in a summer-long procession of death and decomposing. Gradually the dissolved molecular oxygen in these lower waters gives out, followed by anything to which oxygen is attached: nitrate, nitrite, sulfate, finally even carbon dioxide. The order is always the same. In their place noxious gases evolve: ammonia, hydrogen sulfide, more than likely a trace of methane. It is as though the hypolimnion came from another planet, from another Earth-age, from an Earth fecund with the unoxidized molecules of life. To this hidden lake the settling organisms yield up their cellular nitrogen and phosphorus and carbon. The knot that photosynthesis has tied is quietly undone, unraveled in lightlessness, as respiration releases things back into the unformed, into the possible. With no oxygen, the muds begin to decompose, to release manganese and iron and more phosphorus from their locked positions in the once robust but now failing fabric of metal oxides. By August the lower lake is a chemical brew, an exotic, stygian place, lurking just below the casual swimmer's feet. These two lakes, irreconcilable yet interdependent in the drifting, slow exchange of matter, coexist by virtue of their different densities, one curious lake perched above another.

It is in the seasons of a lake that you can sense the miracle of water at work. As summer approaches and the surface warms, the molecules of the liquid quicken their motion. Energetic and vibrant, they spin from confining clusters and, like dancers in the quickening tempo of the dance, carve out more space for themselves, room to turn and pirouette, tiny maneuvers. A warmer parcel of water expands, occupies more space, becomes less dense, a thing of lightness. It fills the surface and isolates below it the cooler waters of the deep. Sheltered from the summer air, from the sun's light, the lower lake lies in its chill repose. Its molecules are held more tightly, held in the thrall of hydrogen bond linked upon hydrogen bond, a vast and sprawling bridgework of more rigid design. These molecules of the lower lake are more compact, more grave and circumscribed in their motion, less given to the wild kinetic flights that higher temperatures occasion in matter. Epilimnion and hypolimnion, so utterly different. And yet, like a volume of Rutherford and a volume of Yeats casually stacked upon a cluttered desk, they are written in a common, endlessly variable language. The atom's emptiness; the wild profusion of Innisfree.

As autumn approaches, there comes a subtle tug toward unity, toward the mingling of waters and the merging of worlds. In the crisp air of late September and early October, and particularly at night, as the stars rise and fall above cloudless skies, the warm surface lake radiates heat away into the autumn air. Its temperature becomes more nearly that of the deep waters. If you run the cable of a temperature probe through the water column from top to bottom, the thin needle barely moves: constant temperature, constant density, all the way down.

In this fine state of balance, a gentle night wind, sloughing the clattering leaves of the maples, is enough to set the poised waters streaming. Masses of stagnant hypolimnion plume upward, mushroom and bend, curl like ancient scrolls through the upper lake. In exchange, the surface waters plunge downward and spread along the fetid sulfidic muds, bringing with them a veritable hailstorm of that fateful and aggressive molecule, oxygen. At fall overturn, the hypolimnion becomes stripped of its methane and sulfides. Encountering oxygen, iron and manganese are oxidized and rain as solids from the lake again, as they must have in ancient seas. The upwelled phosphorus and nitrogen, not to be lost in the economy, spark a fall bloom of algae, a momentary green suffusion on the surface lake. Things enter from below, long stored away; the lake's past comes to light.

Winter stratification in late October seems anticlimax. Beneath the thin veneer of ice that spreads across the surface, the lake is more homogeneous. Today a gray lake stretches flat under gray sky, a dormant sheet, as expressionless as shale. The brilliant canopy of trees has become an outreaching of spidery black limbs, and the drama of overturn has long passed. You can see ice beginning to build out from the docks, in the tentative, molecule-by-molecule way that ice builds, freezing, as ice does, from the surface down.

When I return the lake will be frozen and perfectly still. Like Bonney and Fryxell. Like Vanda. For now, I start the car, turn it south toward home, and leave Acton Lake in its dormancy, hoping that in a few months I will see it again.

When I tell people there are lakes in Antarctica, they think surely I am joking. "Lakes there?" they ask. "How can that be? It's all ice and snow. Penguins running around." Then, when I assure them that it's true, they ask, in a more assertive tone, "But they're frozen, of course?" And I say, "Well, yes, there's ice on the surface, but below there's liquid water, sometimes as deep as two hundred feet." Then they ask—and this is inevitable—"Are there fish?" I say, "No, not a single one." "Hmmm," they respond, incredulous, "a lake without fish. Does anything live in them, at all?" And they emphasize "at all." "Only algae and bacteria," I say. "Nothing you can actually see with your eyes. Except for the mats of algae, which are tiny columns and pinnacles on the bottom, far below the ice."

But then it is precisely what is not there, what has never been there, that makes the lakes—indeed, the whole continent on which they lie—so strange and so important.

For me these absences, and the simplicity to which they give rise, were the key. The lakes are the most isolated inland waters in the world. Landlocked, they are without spillage or outflow; each has only a few streams, and these hold water for only a few weeks out of the year. They are ice-covered, so that very little in the form of dust or snow enters them from the air. And, of course, there is never rain. That in itself makes them magic. How can you have a lake without rain? A lake without fish, maybe, but a lake without rain? A land without rain. A whole continent. Such living things as there are are mostly microscopic—algae, bacteria, yeast, a minimalist's tableau. And into this setting, stark and largely inorganic, Martian almost, the elements come—nitrogen, phosphorus, the metals—unheralded, but replete with possibilities, with lives to be lived.

It would be no exaggeration to say that I was obsessed with the lakes, and especially with the metals that coursed through them like bits and pieces of an invisible wind. In this seemingly fantastical concern, I was not unlike Borges, who once wrote of a silver coin he had dropped into the sea. The coin had become, in consequence, a kind of persona in the drama of the world, its destiny unfolding alongside that of the poet Borges himself. I had my coins, too, by the countless billions.

I knew, for example, that the Onyx River in Wright Valley had brought tons of cobalt and lead and copper into Lake Vanda over its long history. Yet there were virtually no metals in the lake. I knew this. But where had they gone? What was removing them? What thin veil of purity had caught them in its mesh? And whatever veil it was, did it fall elsewhere across the Earth and its seas, purifying as it went? Did the Earth, or this tiny piece of it, regenerate itself? At what speeds? By what agencies? Last year I had set particle "traps" in the lakes, had left them there for a whole year. They were nothing more than clear plastic tubes, capped at one end and suspended below the ice. But in time, if all went well, I would get them back and I would know the answers.

I had hardly slept, had tossed in and out of dream all night. In the dream, winds came down the long valleys, sweeping thin snow before them, turning everything white and opaque, until my hand became ghostly, disappeared before my face. Creaking metal, the movement of giant frames through the air, bending, moaning with the uplift; the boom of canvas, of tent walls filtering the perpetual light. I rose under mountains, in the salt-weathered hollows of boulders, in narrow passes —the continent rising away from me, as it does, a great plain of ice and solitude, of Edwardian figures with their sledges, dragging, barely moving against the hard blue sky. Then I awoke, looked up at the cherry wood of the bed, at the darkness of Ohio, and remembered I was still here. "It's time to pack," Wanda said. "Time to wake the girls so they can see you off." In the basement the furnace rattled as I dressed.

In a few moments I was downstairs. The sun stood large and red on the horizon beyond the water tower, just about to begin its climb through the morning sky. We were all standing at the door—Dana in her yellow sleeper; Kate, who barely came to my waist, in pink. Wanda was dressed in the flowing muumuu she had bought that year in Hawaii. We were hugging and holding back tears and I was feeling that odd mixture of excitement and guilt, even dread that accompanies these journeys. So much seems to fold and entangle itself in this work. So much that is never said. Anticipation of things to come; regret for things missed. How many Christmases do you get with your children when they are still filled with wonder, knowing the winter snows are tossed in magic?

The van from the university drew up to the curb. Mike let the engine idle, jumped out, ran across the lawn, and lifted Katy over his head. Wanda hugged him and said, "You two look after each other. Don't do anything foolish down there. I want to see everyone back here safe and sound." Dana was wiping away tears. "Write me, Daddy. Call on my birthday," she said, "like you did last year. We'll keep the Christmas tree up for you. I promise." "Write to your mother," Wanda said to me. "She seems so concerned this time. And don't forget the shell."

We drove down High Street, jogged right, then left, and up the winding road to Boyd Hall. Walt, Tim, Varner, and Dr. Yu were waiting at the front door, and we all greeted each other with day-of-departure enthusiasm. Upstairs, the laboratory's floor-to-ceiling windows let in a pale light that scattered across the benches and the instruments—the water baths and spectrophotometers, the burettes clamped and covered on the metal stand—and reflected onto the chalk board with its lists and equations and onto the sturdy wooden crates that were neatly lined against the wall. We locked the equipment boxes with their water samplers and filters and pumps and pipettes, and slid the dead weight of the sediment corer into its hinged container. We gathered the black carrying cases that held the pH meters, the oxygen meters, and probes. We stuffed books into knapsacks: *Standard Methods for the Examination of Water and Wastewater; Aquatic Chemistry; Wetzel's Limnology.* Large, ponderous tomes, but for us essentials, tools of the trade.

After we had checked everything against the list on the board, we carried it all downstairs. Mike, Walt, Tim, Dr. Yu, Varner, and I, like a line of porters and sherpas, wound our way down the narrow staircase of the old building, past the brass pendulum that swings in the stairwell, past the polished reading room, and out the door. We nearly filled the back of the van with our supplies. And this was only the beginning. The cold-weather clothing would come later, in New Zealand, and the camping gear and supplies would be carefully chosen in Antarctica, at McMurdo Station. "Expedition," Mike said. "To the valleys of Mars and beyond," Walt continued, raising his arm with a flourish. I turned to look at Boyd Hall. The gray limestone was adorned with a single word carved in large letters above the door:

SCIENCE, it read, almost wistfully, as though somehow there were still only one. We moved slowly up the empty drive that runs by the small Gothic chapel and by Peabody Hall and then drops suddenly down toward the shallow pond, crosses the stone bridge, and climbs back again, past the art gallery and out to the highway. We were under way.

At the airport we unloaded the van, stacked the trunks and cases and knapsacks neatly by the counter, and then walked in aimless little circles while we waited. Brilliant surfaces, reflections, whispers, the serious self-confidence of airports—somehow, in our checkered shirts and hiking boots and faded jeans, our "geochemical attire," as Mike called it, we did not fit.

"Would you please open these boxes, sir?" There was a stern voice speaking behind me. "I hear something. It's ticking." I listened for a second, crouching down, pressing my ear to one of the containers. There was no denying it: *tick, tick, tick, tick,* slightly muffled, but absolutely steady . . . and loud. I felt betrayed. By my own equipment.

A small crowd began to gather as I rooted beneath a mound of pipette tips and filter papers, things that sounded like dry leaves, that flew up and fluttered when I touched them. My hands clutched something cold, round, heavy. People were bending down. The circle of onlookers had grown. "It's the clock drive," I said, pulling it up with both hands. "For the water-level recorder. For the streams." My voice was rising. I was in my own country, speaking English, making perfect sense, at least to myself, and yet I occupied the center of an expanding circle of incomprehension. How to explain this?

I stood up, pulled off the top of the instrument to expose the rounded drum. Varner reached into his knapsack and produced a roll of chart paper. Dr. Yu, who understood intuitively that this was about to become a public demonstration for the "authorities," handed me the yellow float. I attached it to the instrument. You could now see the drum, with its lined chart paper, slowly turning. You could see the pen lean inward against the paper, touch it, leave a trace. You could see the way the yellow float, which dangled at my knees, controlled the pen, moved it up and down against the drum. "Imagine this is a stream," I said, looking out over the crowd, and I pointed down, down at the carpeted airport floor. "When the stream rises, it lifts the float." I lifted it slightly. "And the float lifts the pen." A thin red line shot upward like a spike on the white graph paper. "Like this." A woman in the back smiled. Then smiles all around.

There were more questions about the meters: pH, oxygen, conductivity. They had a look of menace about them, needles drifting mysteriously across the white calibrated landscapes of the faces. I lifted each of them in turn, held them up for the clerk to see, played with the knobs, watched the readings come into view. Readings of nothing. "You can put those things away now. Sorry for the bother. You understand."

▾ ▾ ▾

We were shortly airborne, the plane lifting slowly west, following for a moment the course of the broad Ohio River. I folded my hands on my lap, rolled my head back against the seat, and looked out onto the scattered fleece of morning cloud, onto the plane's detached shadow sliding flat far below across the Earth.

This atmosphere through which we climbed, which lifted us like a great gentle hand, has such a thinness, an insubstantial quality to it. It is little more than a tenuous gathering of small molecules: nitrogen, oxygen, water vapor, argon, carbon dioxide, jostling, colliding, glancing off one another like tiny billiard balls. No strong intermolecular linkages, no hydrogen bonds, no clear polarities as in the water of a lake. Just the weak force of gravity like some invisible shepherd drawing together his flock. A streak, a wisp, a swirl of matter strewn around. With a few bits of data, a few equations, you can count the molecules, the atoms. In the whole atmosphere, there aren't many.

And yet in a way this little is so much. The whole biosphere, the whole tangle and undergrowth of life—the profligate lignins and cellulose of plants, the matlike hemes and porphyrins, the helical proteins winding and unwinding—comes from this, this drifting reservoir of the unjoined and disunited. Our breathing binds us to this air, and thus to each other, to everything living, in a common breath, a common exchange of oxygen, from lung to limb and back again, transformed. The sun's energy, burning on the equator, is sped poleward in giant cells and lariats of air that never weary or cease, that warm the farthest ice-laden sea. This atmosphere connects all with all; with alacrity, it dispatches dew and dust alike. How disastrous it would be if this were not so, if over the great forests oxygen hung in reactive clouds, undispersed and lethal, inviting fire and ruin; or over the cities the vapors settled dark and heavy as rain. But the atmosphere, sweeping, mixing, transferring, never storing very long, will not allow it.

The lower atmosphere, this troposphere, literally the "sphere of turning," is a wild and errant place, a place raging with storms, the fickle, unpredictable weather of our world. Even on a calm day the plane protests, seems to bend and creak. In the troposphere the power of destruction abides, the hurricanes moving massive waters toward the dark shore, the tornadoes capable of razing whole towns: Xenia, Ohio, obliterated, reduced to cornfield rubble in an instant.

But more than this, the troposphere is motion, and transit and life. Imagine a cabin on a winter's day, motes of carbon and steam curling from its chimney. The smoke trails off but never just lies in luxuriant streaks above the Earth. It billows and stretches, dips groundward, rises again and drifts into cloud. The troposphere is an engine of turbulence and change, powered by the fusion furnace of a distant sun. From the heated Earth, warm air rises and on rising cools, then descends again. It is thus that the green kite slips away on the April breeze; the fly ball, helped by an updraft, carries beyond the hapless fielder's leap; the lilies, the grasses, bend on the spring air; the firefly in the folds of the summer yard eludes the child with the jar; the gem of Venus is pushed aside by cloud. In the troposphere we hear the wet trees slap against the attic roof, bearing us far into sleep; and the maple seed glides like a wooden blade in whispers from the parent tree.

Gaining altitude now, we have left all this. We are in the calm near the stratosphere. Here you can see the plane's contrail. The white exhaust hangs, begins to bead like islands, an island chain stretched across blue. Below, rivers and lakes flare like metal and glass, signal mirrors in the brown suede of landscape. Over the Grand Canyon, the pilot dips left and then right, exclaiming on the color of the stone— the reds and reddish browns that iron and oxygen together make. Descending beyond the San Bernardino Mountains, a prairie of housing tracts begins to grow out of sere grass, and blue sky gives way to whiskey, the photochemical hue of Los Angeles, with its aldehydes and ozone, its nitrogen dioxide taint.

On the ground at Los Angeles, before the flight to Hawaii, we drive to the beach, ride strong November waves, toss a football at the surf's lacing edge, stretch ourselves before the long confinement. Near sunset we are back at Los Angeles International Airport. Blue runway lights stretch through dusk to the sea. The silvery planes launch outward on their westward climb over the Pacific, roll in the last pink of the evening sky, like large slow fish, then disappear.

No sooner are we aloft than the continent of cliffs and lights drops into black water and begins to fade. I sip bourbon from among the melting ice, feel the warmth flow from my fingers into the cool glass. Through the gentle haze of ethanol and plain weariness, I am wondering how 1 came to be here, in this 747, above this sea, gazing out on the shifting veils of the evening, heading toward Antarctica.

I cannot exactly trace the path that leads here. Perhaps there is no path, only matted brush, a few indecipherable tracks. Things get lost and memory is always part fiction. But I can say that at some point I found geochemistry and without much plan or forethought it began to occupy much of my time. In geochemistry the chemical elements were not mere symbols on a chart, beautiful as those symbols were—hydrogen, helium, lithium, beryllium, singing out almost as I spoke their names. They were voyagers among rivers and mountains, visitors to the atmosphere, dwellers in the abyss. In geochemistry the elements came to life, propelled by the forces of wind and water and sunlight, constrained and animated by the laws of physics and chemistry. I began to think of them as immortals roaming the planet, tiny gods whose adventures would make a mythologist blush. In geochemistry I heard the biblical voice from long ago:

> *The wind goeth toward the south,*
> *and turneth about unto the north;*
> *it whirleth about continually,*
> *and the wind returneth again according to his circuits.*
> *All the rivers run into the sea;*
> *yet the sea is not full;*
> *unto the place from whence the rivers come,*
> *thither they return again.*

And with the wind and with the rivers go the elements, moving, holding their own counsel, forever reshuffled, enduring, locked into this assemblage a moment, then into that. Never very long anywhere. Yet their intersecting journeys make the world, make it build and fall apart.

I could not work on even the smallest geochemical problem without placing myself in the space of the atoms, as they moved in time, as they traced the great cycles. What would it be like to be an atom of manganese, I wondered. To be iron or copper? To be calcium or lead or carbon? To have time open up to you all at once, like a landscape seen through the opening doors of a country church? How do these destinies, so different from our own, unfold?

With these questions, thoughts of the Antarctic lakes came to me. The lakes were tiny, inconsequential in any global sense. And yet like Blake's grain of sand, they held a world. If you could write the biography of manganese in Lake Vanda, for example, if you could record its history from the time that water first pried it from rock or set it aswirl in the milky dissolving of carbonates, from the slow collapse of ferromanganese minerals, to the time that it came to rest in the prison of lake sediment, then perhaps you could say something, suggest something, about manganese on a grander scale. Perhaps you could say something about its magellenic trek through the world oceans, or its capture in the brown disk of a manganese nodule. Maybe. More than anything else, as I remembered them from those days with Hatcher and Benoit, the Antarctic lakes were tractable; they were geochemical microcosms. They opened onto something larger than themselves.

Hatcher and Benoit. Names that go back nearly twenty years. In some way the journey might have begun then, with them. It was quite by accident that I was living then in a small trailer on the outskirts of town. Sharing it with Hatcher and Hall. My dissertation research on the solubility of gases in molten salts was nearly complete, all except for the writing and I think I was looking for an excuse to postpone that. So when Hatcher told me Benoit was thinking about doing some chemistry in the lakes of the Antarctic Dry Valleys—"Maybe dissolved oxygen, ammonia, sulfide," he said, "things of biological interest. He'll need a chemist. How would you like to go?"—I said, "Sure, I could use a break from the lab, the dissertation. When do we leave?"

We left in August of that year, in the Antarctic spring. The LC-130 Hercules cutting with all its strength into the polar night out beyond the watery, unmarked "point of safe return." We carried enough fuel in a huge bladder so that the plane could circle back, return to New Zealand, if the weather proved too severe for landing. Sometimes in August, you couldn't tell the white mountains of Antarctica from the sky.

There was nothing much at McMurdo then: the building where we slept, a piece of corrugated sheet metal curled over the volcanic ash of the island; the ship's store; the mess, the garage smelling of diesel, the machine shop, and the hangar for the helicopters. Jamesway huts hunkered low to the ground, rounded against the wind, separated far enough so that fire couldn't spread easily. But there were reminders everywhere of Scott: the rambling hut on the peninsula where he had spent part of the winter of 1911. Just walking by I could taste the acrid seal blubber burned in stoves nearly a century ago. Had they been gone only hours, I wondered. Could they still be warned? Called back? But the wooden cross high on Observation Hill spoke of death: "To strive, to seek, to find, and not to yield," it read, facing out in a great embrace toward the Ross Sea, The Barrier, the Valleys, the bodies of Scott and his men—Bowers, Evans, Oates, Wilson—buried out there in fine desert snow. Moving with the moving ice that collapsed into the sea.

In the McMurdo Biolab there was a tiny room that passed for a library. From there I had once put through a phone patch to my parents. They received it in Pittsburgh at three in the morning. It had gone through a ham operator somewhere in Indiana. "Are you all right, Bill?" my father had yelled, his voice cracking and confused with sleep. There was an edge of panic in his words that I had never heard before. "Are you alive?" It took us more than a few minutes just to connect, just for my heart to slow down. And by then he had handed the phone across the bed to my mother. They had never carried on a one-way conversation, with its awful pauses and clicks and "overs." Neither had I. Not through eleven thousand miles.

That year Hatcher, Benoit, and I visited all the lakes. Some in darkness, when you could still see stars in the deep chutes of stone and wind that were the valleys, and sunsets that turned the sky emerald and mauve. There were Russian scientists at Lake Bonney, when the hut was still there, before the lake level rose and they had to take it out. Pinned down by October winds and blowing snow, we talked of Tolstoy and Turgenev, the Petersburg of the czars. There were boxes of frozen steaks, a small oil stove, a Primus, a few plywood bunks, lots of vodka and tobacco. Beyond, in the valleys, there was absolutely no one. There never had been. The nearest things to life were the mummified seals, crawled up from the Ross Sea, bones and flesh still intact. They had been lying there a thousand years almost unchanged.

In those days there were no field radios. They just put you in by helicopter on the frozen lake and said good-bye. When the day for pickup came, you waited. Sometimes you waited for a week until the weather broke, until the winds died down so the choppers could move up valley without being blown right back out. We were living on

K-rations, in tents that were dots of green and red in the landscape. Early in the season—September, October—we could work outside only an hour or less. Even under the hood of the parka, the tip of my nose turned to frozen flesh, white and blistered when I looked at it in the silver blade of my knife. It seemed I could watch my own spit freeze before it hit the ground.

And there was the wind, constant and depressing. It could make you want to hug yourself and rock in the little tent, burrow into warmth and memory. Anything not to have to face it. But sometimes it could be a conspirator in play. On Bonney, we all "surfed" the ice on plywood squares—even Benoit and the Russian geologist Boris Lopatin—using our outstretched arms and our parkas as sails, finally crashing into the opposite shore. "Craaazy capitalists," Boris Lopatin had said, laughing uproariously, holding up a glass of vodka whose surface had frozen solid.

I kept a small journal that year. On September 15, there was an eighty-knot wind pouring over the snout of the Taylor Glacier, down the east-west axis of Lake Bonney, slamming into the wooden hut where we slept. The hut rocked back and forth, moaned, uprooted two of its steel wires, and seemed on the verge of being tumbled down the valley. But then the wind suddenly stopped. According to my journal, I got up, walked outside over the scree slope and down onto the lake ice. It was forty below zero. The sky was a cloudless blue and beneath my feet the prismatic ice, cut crystal, was ten feet thick. You could drive a locomotive across it if you wanted. I chopped some of it with an axe, carried it back to the cabin in my gloved hands, put it in a large pot. I melted it, boiled it, made coffee in a metal cup, and drank it with my gloves on. Then I went out again. There was a granite boulder the size of a person kneeling. It was shaped like the Virgin, hollowed that way, a hood of stone around her smooth oval face. Over coffee, I stared at it, wondering how long it takes wind to carve rock that way, to abrade it grain by grain into a figure in prayer. A thousand years? A million? Time lay all about me, visible in the naked stone.

Suddenly there are strings of light turning in the black water below. The plane rolls and they disappear, then rolls again and they return, bright now, the city of Honolulu glittering in the midnight sea.

In the terminal the dense sweetness of flower leis is all around. The women in their red and blue muumuus drift as in a dream, slow, soundless night motion, the torpor of islands. Far beyond the airport, lighted houses and street lamps climb the flanks of the Koolau Range, leaving in black cleavage the unbuilt ravines and the upper wooded slopes of the mountains.

Gazing up, I can imagine Keith's place, a single light on the farthest ridge, overlooking the volcano. I spent a year working with Keith: the endless talk about the sea, about the vents, the pillows of lava that move out there in the dark, their chemistry. We talked about how the oceans formed, how the islands themselves came to be: the seafloor spreading beneath us, the fire of the mantle welling up and moving away, cooling, leaving its magnetic message in the frozen stone, the plates always shifting over the broken shell of the Earth.

The islands came from all this tectonic motion, this eternal building from within. The Hawaiian myths spoke of it, spoke of the goddess Pele, the goddess of fire spilling lava, raising land, vast mountains domed and hissing above the cold sea. "Born was Pele, in the night," the creation chant said. "Born was the coral, the mother of pearl, the shellfish, the conch." The chant spoke of fire and water and calcareous stone. The new theory of plate tectonics embellished upon this, saw it all unfold in time, gave it dates and causes, tied it to the larger frame of the Earth. The geophysicist Tuzo Wilson argued that the Pacific Plate was gliding above a hot spot in the mantle. As the plate passed, fresh lava welled up from the "spot." It was streaked red and orange with heat along the seafloor, cooling, building its way higher. Several million years ago land finally emerged, struggling against fog and rain and cutting surf. You can see the track of the Pacific Plate, its northwest drift, in the line the islands make in the empty sea.

In the years that these islands of basalt have waited, all things have come. In the beginning there was just rock and steam, Pele's struggle against the tide. Now these outcrops are full. The soft air, the smell of mango, papaya, and ginger, the rustle of palm leaves, the glow of the night sky over the Koolaus, up by Keith's house, tell you this. For these islands have been port to any spore that drifts or floats on tropospheric winds, to any passenger come on the pinions of brilliant birds. It is deep green maidenhair up and down the creased mountains, the petals of flowers adrift, perfume far into the desert night of the sea, the islands in nacre garlanded, the bays and inlets, the broken mouths of volcanoes aswarm with the lightning of sudden fish. I want to stand here and take all of this in, let it bathe me like a fine island mist, this spectacle of matter, the golden elements, concentrated on these shores. Knowing what is soon to be, knowing that in a few days, a few weeks, all of this will hardly seem possible—islands of burgeoning and blooming, the scent of living things compressed, of moonlight falling silver over it all—I question my own footsteps. I question my movement to the plane out on the tarmac waiting.

It was after midnight when we took off. From the window, the lights of Waikiki rose as a thin necklace at the throat of Diamond Head. Soon only the wing was visible, as gray as an industrial rooftop, cutting the palpable blackness. I glanced around the plane. Behind Mike, Dr. Yu, and me, seated in the vast midsection, 1 could see Walt, Tim, and Varner. Walt was still awake. Peter Medawar's *Advice to a Young Scientist* lay facedown on his lap. He seemed content with his decision to come here, to "do science," as he put it.

It had not always been this way. I had expected Walt would go on to study literature or philosophy, aesthetics maybe, somewhere on the West Coast. I think it was the fieldwork that drew him in and made him change his mind: The sampling on Acton Lake, the stream measurements he had made on phosphorus. The lake and the land, they were connected in subtle ways, ways he hadn't thought about. The lake reflected the land, summed it up, told its story through time.

In lakes you can see things happen. Somehow they are just the right scale: not so large that change is imperceptible, not so small that it can be dismissed as merely local. One summer the arcing leaps of the sturgeon are less common. Then the cisco and sauger pike, the white fish and blue pike begin to disappear. Someone notices that the dissolved oxygen is lower. Then the decline accelerates. The mayfly nymphs, once so abundant in the sediments, turn up dead, and in their place bloodworms and fingernail clams abound.

I remember having a conversation like this once with my father. It was under the soft lights of a ballpark, the grass partly in shadow. He was standing there with his arms folded, looking off toward the outfield stanchions and the absolute darkness that lay beyond the fence. Out where the waters of Erie began. "Where are the mayflies this year?" my father said. "I've never seen so few. No clouds of them out there around the lights." He sounded surprised. For a second he was no longer watching the game. I didn't know what he was thinking, but he sounded very sad. *A few missing mayflies*, I thought. *What does it matter?* But I said nothing.

Like my father, Walt knew about these things almost intuitively. He knew how a lake could change, what it could say, what it could tell us about the way we lived on this Earth. When he saw that, intuited it in his numbers, it was too much. He was hooked.

For Tim, things had been clear now for some time. He had discovered analytical chemistry, had taken to the lab and all of its delicate operations with a deftness and good cheer that I had never seen before. His work had a kind of meditative stillness about it. A saintly quality. As though every act were being offered up. You hardly knew he was there, and yet everything got done. Meticulously. Almost without effort. Nothing wasted. I would just scratch my head and say, "How did he do that? So quickly?" I knew every number was exactly right. And with Tim there was something else: years of camping and fishing with his father in Canada had given him the resourcefulness of the outdoorsman. He could probably make a canoe with tree bark and fish with carved stones.

Then there was Mike, black hair down to his shoulders, a slight roll to his walk, that dense thicket of beard like Raphael's Aristotle. Mike, alone of all of us, you could imagine on the first polar expeditions with Scott or Shackleton, always indefatigable and good-humored, ready to put his life on the line. Varner called him a gentle giant, even though he stood only five feet ten.

Dr. Yu, the oldest in the group, was a visiting scientist from Qinghai Province in China. He knew geochemistry, wrote poetry about the high, barren, windy emptiness of the Tibet Plateau, and was an expert on the salt lakes of that region. He would understand the Dry Valleys instinctively.

Varner was different. From the time I met him, he seemed a man of reason. Not easily excited. A skeptic, cautious. Even now he looked uncertain about our project. Maybe the flume he was designing to measure stream flow would work, maybe it wouldn't. He would give it his best. But he could not even imagine the valleys and the mountains of Antarctica, he said. He could not imagine the streams or his own science there. The glaciers seemed like absolute fictions. I think, secretly, he thought I was slightly mad in wanting to go back to the lakes year after year. "Once should be enough," he said. "Maybe more than enough." Sometimes I think he doubted there were lakes there at all.

It still seemed odd that he had decided to come. We had been friends for many years. In college we had taken chemistry together, had run through the infamous "qual scheme" in laboratories that were stifling and dark, that smelled of rotten eggs. "So this is chemistry," Varner had said wryly, shaking his head, as though some life decision were being made right there on the spot. That year we heard the great chemist Linus Pauling speak of disarmament and peace and then glide, almost without pause, into the theory of the hydrogen bond. "It was recognized some decades ago," Pauling began, "that under certain conditions an atom of hydrogen is attracted by rather strong forces to two atoms, instead of only one, so that it may be considered to be acting as a bond between them." Then he stopped for a second, as he often did for dramatic effect, and then, with his arms spread wide, his face lifted to the crowd, he said, "This is the hydrogen bond." During the hour, he spoke of the properties of liquids, the structure of proteins, the origin of life, all the while returning to the little bond with which he had begun, the bond that influenced everything. "A man of passion, brilliant," Varner had said without emotion, as though he were offering a pure description.

A chemical engineer by training, Varner had once earned a big salary blending polymers, turning plastics into corporate gold. But a decade into his chosen career, he began to have doubts. He went back to school at nights, became a high school physics teacher in Akron, working for less than a third of his engineer's pay. No regrets. He was living a good life in Akron, in the big restored house he had bought eighteen years ago, not far from the Firestone mansion. He had his backyard garden, with its early tomatoes, and the barn swallows nested nearby. Varner was a putterer, a repairman, a man of the neighborhood. But he was also a gifted engineer. If anyone could turn a piece of plywood into a flume and have that flume get numbers, it was Varner. It really didn't matter that he was not a hydrologist and had probably never even taken a course in hydrology. He had the engineer's "feel" for physical objects, for devices and gizmos, for things that made the world turn. So Varner would build us a flume, tell us how the streams ran in time.

How the streams ran was at the heart of our work. We wanted to know how the streams gave rise to the lakes, how the lakes were prefigured in the waters of the streams. And we wanted to know what the streams carried, what dissolved things swam like secrets in the current:

how much manganese and iron and cobalt, how much phosphorus and nitrogen, how much calcium? If you could measure the streams you could track the journeys of the elements, imagine them in the small compass of the lake like migratory birds.

As the project leader, I had parceled out the work according to interest and skill, but mostly according to interest. The fire had to be there first; the other would follow. Walt would study phosphorus and nitrogen, would try to measure how much of these elements the clear glacial melt streams brought to the lakes each year. We were interested especially in phosphorus, because in lakes this is the element of life, and sometimes of death. Mike would look at how the lakes had evolved through time, at their chemistry and salt structures, at how quickly the streams brought calcium and magnesium and sodium and potassium to them. Chemically the lakes were all different. Even when they lay cheek by jowl in the same valley, they were different. Why, we wondered. To explain the lakes we had to understand the streams. We had to know how fast they flowed, how quickly they delivered their precious cargo of ions and salts and clays. Getting the flows right was Varner's job, and so much depended on it.

And then we had to retrieve the particle traps from Lake Vanda—this was up to Tim and Dr. Yu—gather together what bits of matter had settled into them, and decipher that. I worried about the traps. We had left them suspended for a year, tubes dangling at the end of a thread. Yet if all went well there would be signs and messages in them. For while the streams spoke of arrivals, of the elements coming down to the lake, the particles spoke of departures, of things leaving. In the two there should be balance.

▼ ▼ ▼

A deep voice from the cabin interrupted my last-minute audit. We were crossing the equator now, it said, and I glanced out the window as though there were actually something to see: A thin blue line, perhaps, like the one they pasted on the spyglasses of nineteenth-century sailors when they first came into these parts, or even a few soft zeros marking the mystical latitude of the Earth's halving.

What had brought me here, then, hurtling over the dark Earth, heading south? As with every journey, this one began with a few conscious steps, and I could record those. They have dates and dollar amounts and postage. But they are only a small part of the whole. *Time and chance happeneth to us all, saith the Preacher.* So what is it that comes before the easily recollected numbers, the presence of certainty? What lies in a time before Benoit and Hatcher, before I had even heard of the continent? I might just as well have laughed at the thought of this southern extreme the day Hatcher first mentioned it. But I didn't and I wonder why. What did the word *Antarctica* reach into and touch? What resonances did it stir in what had long been stored away? Things I didn't know had been there to begin with.

I went back to sipping a Jim Beam and searching for some visible evidence of our half-world position—something physical besides the wing light's pulsing with the odd glow that it cast before being swallowed up. You had to laugh at that. A whole universe filled with night and then the winglight, like a little prayer. But there was nothing there. Only the rising winds. And far below, the empty sea.

TWO

Water

Our house stood in a raw little valley, almost a hollow, heavy with work, the night shifts and the day shifts blending in dim bars. Men trading stories of the mills. Beyond the windows the hills rose up and attached themselves to the sky like green-striped awnings. The steep, narrow streets switched back on themselves, hairpin turns of red brick spiraling up. Visibility was a mile this way, a mile that way, the horizon always close, cluttered with chimneys, dark coal smoke in winter from the anthracite. I must have been seven when I first left this place for the sea. I think the ocean was the first real openness I had ever seen. It was the year after we had gone to the forest, when I felt so closed in by dark pines I could hardly breathe. "You'll love the ocean," my father said. "You can't see to the other side."

He found us a clapboard rooming house, the paint peeling and blistered from the salt air. You could taste the ions on your tongue, feel their charges as a rush of exhilaration that came out of nowhere. We unpacked in a tiny room and headed for the beach—Eugene and Elizabeth and I following closely in our father's wake. The sky was brilliant, a few white clouds patterned above the sea. As we approached the boardwalk, I fell a step behind. I think I was afraid of what I would see, that it would not measure up. The planks steamed, reflected mirror light. I looked over the worn wood that was old and smooth, almost like glass. There was a dazzling expanse of white sand before me, trembling thin lines of surf the length of the world. Everything moved or seemed to move, to break into hot points of radiance. A billion coruscations dancing in the foreground of the sea. Everywhere water, the heaviness of light. My father was out there, barrel-chested, booming against the waves, his white T-shirt, which he never removed, fluttering around his waist. I just stood looking at it all, the glittering openness that went on and on, that opened my breath, watching his large pale arm wave us in.

▼ ▼ ▼

WATER IS SUCH STRANGE STUFF—so ubiquitous we hardly see it. It might as well be invisible, like one of those weightless fluids that followed Newton into the Age of Reason: phlogiston, caloric, the luminiferous ether, all transparent and aflutter with light. I walk on any ocean beach and the sea takes me in, almost as a friend, a fellow creature, a member of the living clan of shapes and sizes and movement that stretches hopelessly deep into its profligate past. We seem to know one another, this sea and I. I enter, and it subsumes me into itself. I suppose I am nothing more than a thin jelly, a flood of cells, water sluicing in veins and pipes, nothing it has not seen before. It cannot distinguish where I begin and it leaves off. It senses me only as one of its own.

Indeed, we are watery reeds in a watery world. So much water bound in watery sacks. Even our minds and our very breath rise and fall on the watery tide: eighty percent of our brains and our lungs, sixty-five percent of our entire bodies. Water courses through us, bathes each cell. The moon tugs at these inner seas.

The Earth, too, is irrigated by water: seventy-one percent of its surface, 320 million cubic miles of it. And there is nearly half this much tied up as "water of hydration," bound to the ions of solids in the Earth's crust. An invisible sea. Once you leave Los Angeles on this journey south to the Pole, there is nothing but water—all the way to Antarctica. Perhaps there are a few brown island stones, as smooth as skulls down there, rising above the waves. But even the twin islands of New Zealand seem little more than Maori canoes in the vast Pacific. Beyond Christchurch there is nothing to speak of either but water, until you cross over Wilkes Land and Cape Adare and encounter . . . yes, more water, though now in the form of Antarctic ice, as far in every direction as the eye can see. (Stephen Pyne has said of "Greater Antarctica" that this is "a world derived from a single substance— water.") Twenty hours aloft over the Pacific and you get some sense of what the surface Earth is really all about. Solid Ohio, with its browns and greens, its pastures and fences, seems little more than a cartographer's fiction of soils, plants, and trees. From the observatories of space, from the dry moon itself, it is the signature and brushstroke of water that you see, a delicate blue wash across the Earth.

Most of this water is in the sea, of course, mixed there with the weathered salts of the continents as churning brine. The remaining fraction is fresh, but most of this is sequestered in ice caps, hoarded there by darkness and cold. What is not stored at the Poles occupies the more familiar niches of lakes and rivers, atmosphere and ground-water. The Antarctic ice cap alone is so vast a storehouse of water that if melted, it would charge the rivers of the world for more than eight hundred years. And the Gulf Stream, wide and sluggish, bringing improbable springtime flowers to fogbound New England coasts, carries in its flow twenty-five times as much water as all the Amazons, Congos, and Mississippis on Earth. When these numbers are taken in and sorted out, it would seem that the hydrosphere—the realm of water in all of its pooled and dropleted and vaporous and streaming forms—was prepared and apportioned more for pilot whales and penguins, anemones and seals, than for those of us who dwell upon land. "Make no mistake," Varner once said, "water is ubiquitous and yet it is more precious than gold."

And more magical. There is a delicacy and power to this fluid that rushes through our veins and our lives, that punctuates waking and dream with longing, whose pools are light among blowing sand, among dark forests and low hills, whose forms—solid, liquid, vapor—are legion. I remember the snows of Pittsburgh, how the flakes grew as they fell, how they thickened and slowed in the viscous air, how they built dangerously on the rooftops until the roofs creaked and bowed under the weight, how mountains broke and avalanched under the winter trees. Or much later, how the waves, bottle green, curled luminous and lifting over the coral shore. How I went down and could not come up under their load. Water is mass and power: its drop-by-drop accretion into flood, its sudden rush through the breached levee, the brown swarm of its carried silt whispering through the drowned corn.

Once, when the trees of Ohio were silver in morning, they spoke under the gentlest breeze, the clatter of their thin branches ice-cumbered and dazzling, red and green and blue, the spectrum laid out in winter sun. Whole trees with leaves of ice. Whole orchards ablaze in refracted light. When ice forms on a lake, the crystals shoot outward from the cold shore into the center, long fingers extended into pale November afternoons. Dew fixes on the surface of grass, beads in translucent spheres and shakes its silver skin. In the mountains of Montana they said three feet of snow had dropped in a single night. The trains stopped, the cars stopped, the passengers forgot time in rustic inns, sipped hot drinks by the window. The wide streets beyond the parted curtains became legends. Once in Arizona they said the dam might break, so much rain had fallen. Cloudburst after cloudburst. The whole town emptied, the cars moving slowly in lighted procession to high, safe ground. There were just a few of us left on the bar stools, singing, hiding our fear in drink and laughter. But the next day when I awakened there was sun. The waters had flowed off into the night, peaceful and calm, rolling away through the hot desert down to the sea. Skiing on the Great Divide, I saw my daughter Kate stop and shake snow from her left hand. "To the Pacific," she said, laughing. And from her right hand: "To the Atlantic." A great smile lit her face.

Water is everywhere, we just need to remember it. It is in the trees—hundreds of liters transpire each day from an ordinary elm, heaved as gentle fog into the sky. There are a billion billion molecules in a single flake of snow, which is why the great photographer W. A. Bentley, the "snowflake man," could say that in forty-five years he had never seen two alike. And never would. Water is imbibed in the germination of seeds, a process that involves the uptake of water molecules by coils of cellulose and loops of starch. This "drinking in" can unleash tremendous forces; thus small seeds split rocks weighing tons. Water imbibed by wood can create a force a thousand times atmospheric pressure, which is why the Egyptians used wetted stakes to split limestone for the Pyramids.

The early Greeks thought that water was not only everywhere, it was everything: every material object, every stone and cloud, every citizen of Crete, was nothing more than water transformed. One substance per world. What economy! Thales of Miletus asked, "What is it that changes, what is it that lies beneath all that we see and feel." Things come and go, wind turning to rain, rain to earth, earth to sea, the cycles like Ephesian winds blowing on forever. And yet something endures in all of this. Something is merely changing form. And Thales answered, "It is water."

Philosophers say that Thales posed one of the truly great questions ever about the world. It was the question that began the quest —the quest whose conduct has affected us all, whose revelations and outcomes we still await. It is worth considering for a moment. Already in the sixth century B.C., a mind was driven to find unity in the "manifold of phenomena." As a start, Thales proposed a kind of "primary matter" whose permutations and combinations, whose rearrangements in space, could account for all that we see. That he should have chosen water seems a triumph of observation. For it appears for all the world that water, when it freezes, can become stone; and when it vaporizes, rises into steam-white clouds, can become air; and when it flows to the sea, can yield up earth in trellises of dark loam as deltas; and when it appears as storm, can give rise to the skittering fire of lightning. So many forms and so many names.

Water is everywhere, so we take it happily for granted, think it somehow usual that rivers should drain the planet's skin, that seas should lie in its basins and folds. That rain should come from the skies along with snow and mist and dew. But we know thousands of liquids. Tens of thousands. Among all of these, water is singular, almost preternaturally strange.

At what temperature, for example, should water boil? If we knew only the molecular siblings of water—hydrogen sulfide, hydrogen selenide, hydrogen telluride, all of which look like water when you write their formulas down—then we would expect water to boil near minus eighty degrees Celsius. The kettle would sing in the dark, frozen night of the world. If water behaved like its siblings, it would be a gas at Earth temperatures and the atmosphere would roll with its troubled cloud banks and the sea would be a hovering fog.

Even as a staid and proper solid, water is no more predictable. On a winter lake, ice and zero-degree Celsius water move atop the denser liquid below. There is nothing dramatic in this, or so it seems, until you think just how absolutely strange it is that solid ice should float. The way of the world is for solids to sink in their own liquids; for cooling lead or mercury or methanol to settle out of their own fluids and crystallize. Thus things freeze from the bottom up, the molecules of the solid becoming neatly packed, as dense as cannonballs on a town green. But ice is different. As it forms from the surrounding water, the molecules open outward, link in delicate lattices and structures, geodesic domes buoyant on the water below. Because of this simple, immensely complex fact, the winter lake lies protected by the solid phase of its own self, its heat sealed and safe from the chilling air above. Because of this fact, fish inhabit northern lakes.

The buoyancy of ice is borne of its expansion. Other substances, regardless of their state, shrink as they cool, and for the most part, it is true of water too. But below about four degrees Celsius, something quite unexpected happens. As the temperature edges further downward toward freezing, liquid water begins, almost incredibly, to distend. If we were to view its cooling in a fine tube, as the Florentine scholars of the seventeenth century did, we would be amazed to see the shrinking column reverse its travel and begin to move steadily upward. This expansion of water continues as the temperature is lowered to zero. Finally, as water freezes into solid ice, its volume increases even more. The increase is dramatic—a full nine percent over the volume of the liquid. I have seen water, frozen in a confined space, shatter steel, embed shards of metal in laboratory walls. And the very same thing happens when drops of water freeze in winter stone. The water expands, the stone breaks, and slowly the entire mountain falls.

So, if water behaved like a normal liquid, there would be no lakes in the Antarctic Dry Valleys, only blocks of ice. Water would solidify first at the bottom, along the sediments, as needles and fingers of ice. Every living cell would be gradually locked into stillness, into an eternity of cold. But as it happens, ice freezes and floats on the surface. The waters below are protected from the fierce winter night as though they were cosseted in wool. And beneath this ice, in the liquid water that moves below, things can live and much can happen.

Yet the anomalies of water are not limited to cold temperatures. With the exception of ammonia, water has the highest heat capacity of any liquid or solid on Earth. This means that it has a kind of thermal inertia—its temperature is not easily moved by heat. "A watched pot never boils," we mutter impatiently. And from childhood we remember that the sidewalk puddle is always deliciously cooler than the scorched summer pavement on which it sits. Because water is so slow to warm (and to cool), it has a moderating effect on climate. Cities on the ocean enjoy cool summer breezes from the sea and mild, snowless winters. Surrounded on three sides by water, San Francisco has one of the most equable climates on the continent. In fact, from this single phenomenon comes the relative mildness and uniformity of much of our planet. The oceans store and distribute vast quantities of heat, shift it from the dazzling latitudes of the equator, move it north and south against the rims of continents; an ocean current a hundred miles wide can transport as much heat in a single hour as can be gotten from nearly 200 million tons of coal.

There are other properties of water with Earth-wide consequences. Unlike gases, which have a certain "springiness" to them, liquids can be compressed only with great difficulty. But of all liquids, water is the most compressible, its volume the most responsive to high pressure. And while this effect is slight and is difficult to measure with any but the finest laboratory instruments, its significance on a global scale is considerable. If the volume of water did not shrink at all under pressure, if water were truly incompressible, sea level would be sixty meters higher than it is today, and the land area of the Earth would be reduced by a full five percent. The Netherlands, the towers of Manhattan, all of Bangladesh, and the Maldives would be gone.

Even the very surface of water is unusual. It is the wet, stretched skin of a drum, and the evidence is all around us: A water strider darts across the silver-satin finish of a pond, creating only dimples on the unbroken fabric as it goes. A steel needle, gently placed, floats in a cup of afternoon tea and the higher density of steel is defied. The inept diver breaks the surface of a lake and for a split second there is a lash across the belly, a fierce sting, a redness that will not abate. It is this tautness, this same surface tension that coaxes water into droplets and spheres. The shape of falling rain, the shape of rain on windows, the shape of dew on a blade of grass, on a spear of barley—these round-nesses go back to that surface.

So too does water's power to erode. A drop of water is a bullet fired at the Earth. It is a hard pellet with a hard skin. It can blast tiny fragments from the most solid rock. Rain breaks on the face of the mountain, shatters microlayers of stone, craters the stone with its force, carries it away as sand and silt. All night long the rain falls as hard as sand against the stone, and the stone, in time, disappears.

But among the extraordinary properties of water, perhaps none is more important than its power to dissolve whatever it touches. As if in answer to the alchemist's prayer, water is indeed the universal solvent. It takes everything, to some smaller or larger extent, into its bulk and substance. Oxygen, nitrogen, and carbon dioxide from the air; calcium, magnesium, sodium, and potassium from the stone. The entire periodic table runs with water in every river and rill. The atmosphere lives in it as in a mirror. For these elements, water is the gathering place, the medium of their collusion, their joining and condensation, their prebiotic building into amino acids and proteins, into sacks of living matter. Because it can dissolve, it is the font of life. Because it can dissolve, it can move mountains.

A drop of water touches a crystal of salt. How common a thing this is. For a split second nothing happens: the crystal is sturdy and built to last. To melt table salt, to rattle the cubic cages its ions form, requires high temperatures, foundry temperatures, eight hundred degrees

Celsius. Then, before you realize it, the salt is gone. Dissolved. Become part of a solution, no longer visible as a translucent cube. How, in contact with water, does sodium chloride glide into seeming non-being so offhandedly, with such insouciance and at such modest temperatures that the whole vanishing act appears to be pure presto-chango, solid substance one moment, then gone the next? Change of this order should require extremes: large objects dropping from the sky; blaring trumpets; blast furnaces. Grinning devils with pitchforks, at least. Surely something more respectably intense than cold water.

But there is this trembling at the heart of matter. It goes on and on. Like the molecules of water themselves, the sodium and chloride ions know no peace, chatter constantly at their moorings, yearn like the ships of Greece to be off and wandering. When a drop of water presses against the crystal face, it is as though a great tide had washed ashore. The ions are loosened and borne away, swept up in a flurry of molecules whose charged ends are turned just right for the task. The solid dissolves into invisible navies of charge, dissipates, diffuses, becomes a homogeneous mixture against which the forces of gravity cannot prevail.

Yet all of this talk about physical properties says nothing of water's dynamism, its splendid life upon the Earth—its swift turning in storm and fog, its movement through rivers. No matter where it is, water is moving. It is always in passage, wanting to be elsewhere. Thus water evaporates from the sea, blows over the lands, hovers, condenses into spheres, falls, evaporates again, or trickles downward through the Earth into waiting reservoirs deep among the folded rocks; or runs off the land and back into the sea. We know the times, too, of water's passage. We have measured these things: nine days in the atmosphere, a few years in lakes; a few hundred years in groundwater; a few thousand years in oceans; more than ten thousand years in the ice cap of Antarctica. Always moving. Yearning to be somewhere else.

We have a million names for water in passage, so many forms does it take. The forms are a kind of poem, a line through our lives: lake, river, stream, cataract, waterfall, ground fog, ice, shower, downpour, meander, deluge, spring, font, runnel, marsh, bog, fen, inlet, bight, bay, wave, billow, and swell. And many more. It is like a chant, each word a memory.

It is no wonder people have given their lives to water, to be on it, to use it, to take from its bounty. Sea captains, farmers, shell fishers, river pilots, singing gondoliers, painters, and poets must all know its moods and its ways. No less the limnologist, the geochemist, the oceanographer, the hydrologist, the engineer, the solution chemist, the teacher, and the writer.

No less N. Ernest Dorsey of the U.S. Bureau of Standards and Measures, who provides us, in 1940, with more than six hundred pages of data, equations, tables, and graphs. All on water. Much of it curious and fascinating. In one of Dorsey's entries we find that two inches of ice will be strong enough to hold "a man or a properly spaced infantry," and that four inches will support a horse and rider, as well as light guns. Ten inches of ice is sufficient to accommodate "an army, an innumerable multitude," and fifteen inches will suffice to hold railroad trains and tracks. One astounding entry claims that "twenty-four inches of ice once withstood the impact of a loaded railroad passenger car falling sixty feet through the air," but, alas, broke under the impact of the locomotive and tender.

Some of Dorsey's most intriguing information is on color. Lord Rayleigh believed that the blueness of natural waters was most likely due to the reflection of the sky. But a colleague of Darwin, J. Y. Buchanan, took exception to this opinion, noting that "when quiet water, as in the screw well of the research vessel *Challenger*, is viewed vertically under such conditions as to exclude reflected sky, it appears to be of a beautiful, dark blue color." In 1936, E. Petit solved the mystery when he determined that the blue of Crater Lake, Oregon, arises from light scattered by the water itself. But the most convincing report was written by Charles W. Beebe on his bathysphere descents into the open sea. At fourteen hundred feet, long after the blues of the sky had been lost, Beebe wrote exultantly that the "outside world was, however, a solid, blue-black world, one of which seemed born of a single vibration—blue, blue, forever and forever blue."

There is surely enough beauty in the mere facts of the world. They are ten times over and more, a surfeit. Enough to go round and round. To burden us forever with delight. And yet, even breathless with this, we somehow need more. We must have more. Perhaps it is our tragedy, perhaps our hope. Whatever, we must know why, we must know what is behind it all. Behind the snowflake, the water strider's dance, the lake steaming at dawn, the looming iceberg on the sea scrolled with death—just the sea itself, its very presence and being, that it is here on Earth at all, that it speaks, that its brilliantined surface blinds. And the life of the sea, life in reeds and tubes, a dense matting around the Earth—what unites these? The world is not just pieces and fragments, not just glowing embers of fact, surfaces, baubles and trinkets and gold doubloons, marvelous though these surely are. It goes down and down, deep into dark wells below our seeing, into mystery and dream, then opens in chambers of startling light. Thales knew this. And Lucretius. And Dalton. And Bohr. But who are we, that we must enter these secret places far below the surface of things? The god of the wind, the god of the mountain, the god of the crops, the god of the light-filled sea. Who is the God behind these, the God from whom all these others come?

THREE

▼

Rutherford's Den

Matter sings. In its spinning and tumbling, its locked vibrations, its translatory leaps, it sings, but we cannot hear. Beneath the most placid surface—a water drop from Acton Lake, a table at Christchurch, the glacial erratic on the shores of Vanda—there is a ceaseless, sibilant whispering, a kind of delicate rustling and turning, unattended sounds so profligate and spendthrift, so seductive, that if they did not lie forever beyond us, we would be held totally in their thrall.

▼ ▼ ▼

THREE HOURS OUT OF AUCKLAND the sun rose. The clouds were columns and cliffs in a three-dimensional sky. In time the North Island appeared, sparse in settlement, the bays of Auckland flecked with sails against the brown headlands. We had been more than twenty hours in the air. Had passed from winter into summer, and back again to spring. We had crossed the equator, the dateline near Fiji, had moved ahead an entire day of calendar time. When we landed, there was such relief, such loud applause. It went on for nearly a minute.

We left for Christchurch in early afternoon. The day was clear, and New Zealand lay below, a land of scattered lakes and glaciers, forests and bays. In Christchurch we stayed at the Windsor Hotel, a small bed-and-breakfast in Armagh Street. Old, but open and airy and smelling of spring, it faced onto a large public garden. The curtains of my window billowed in the afternoon breeze and sunlight fell through onto the bed.

I unpacked and went outside: the scent of franjipani, the poinsettias in window boxes against the green lawns. We were in the southern hemisphere now, far away from the sleet and cold of Ohio. We were where water drains clockwise, where it's mostly ocean, where you can feel the Earth tilt inward toward the sun, leaning you obliquely into the face of its warmth instead of away. I hadn't been in Christ-church for more than an hour, and I felt as though I had been living there a decade, an expatriate. It was easy: lounging on the park grass in short sleeves; queuing up at the street vendor's two-wheeled cart for ice cream; sitting on the banks of the Avon River where it wends under poplar and willow, my feet dangling in clear water; sitting in the town square listening to the Wizard, in his starry cape, preach to the pigeons. Even the marble statue of Robert Falcon Scott, in white tunic and polar gloves, looked warm and content.

I did not know how long we would be in New Zealand. Maybe a day, a week. You had to wait until the "Herc," the LC-130 Hercules, was ready to fly, until the weather at McMurdo was good for landing. Sometimes you'd leave Christchurch, fly for a few hours, then turn around. They never told you why, although there were always rumors. I was anxious to get to the lakes now. Here time moved so slowly. There it would just race, carry me along like a swollen stream.

▼ ▼ ▼

Our flight had been moved back indefinitely. There were storms over Ross Island: whiteout and seventy-knot winds. Williams Field was closed except for emergencies. Word came that there had been a terrible crash at Siple, out at the recovery site. One of the LC-130s had hit a crevasse, had jackknifed over its nose onto its back. There were rumors of deaths and of many injuries. You could feel a heavi-ness in the air every time you stopped by the clothing warehouse. People almost whispered when they talked. We were put on alert, advised to stay close to our hotels, to call in every hour, just in case the weather broke.

Morning. There was no plan of action. We fanned out around the city, drank coffee in little shops, ate thin British sandwiches and sausage rolls, wandered the banks of the Avon under its big, shadowy trees. Ripeness. Fullness. Eternal spring. Buses swaying through streets that looked like London. All in the midst of the Pacific. "I could get used to this," Varner said. "Buy some sheep, settle in, make bulky sweaters."

To mark time, we decided to go to the formal gardens and to the Ernest Rutherford exhibit, or, as they call it here, "Rutherford's Den." Now that we might have a few extra days, I wanted to think about Rutherford, this man from New Zealand; I wanted to think about how he had hollowed out the world, taken the measure of its emptiness; how, like Henry Moore, he had urged us to think of space as much as of substance; how he had turned substance into space. "This is where it all began," Varner was saying, as he looked into the tiny exhibit room where the tall Rutherford was holding aloft an intricate glass tube. "This is where we really begin to understand."

There was a day not long ago when matter was thought to be solid and impenetrable, just as it appears. This table to which we have moved in the sunny garden, this table on which I am leaning, supports my coffee cup, my writing pad, my elbow. It supports Varner's journal, which is lying open and empty before us. Like Sir Arthur Eddington's famous table,* this one is everything we could hope for: it presses upward against our arms, remains flat and level against the shifts of weight we impose upon it. It is the perfect surface: opaque, rigid, a thing to be counted on. A fine material object.

The English chemist John Dalton would have recognized this table. He would have recognized its properties as mere extensions of the tiny atoms he had been thinking about for years, atoms that were round and massive and hard. There was nothing smaller than Dalton's atom, and inside there was no space. Like clay spheres in a clay ring, they had no moving parts. As the Greek *atmos* implied, they were the "uncut," and, for Dalton, the uncuttable. Solidity, the solidity of this table, was nothing less than one should expect, given the robust atoms of which it is composed.

The world of objects, at its unseen heart, was a collection of microscopic spheres, variously configured bound into aggregates. These spheres had one fundamental and ultimately knowable property, and that was weight. The passion of Daltonian chemistry was the passion for atomic weights.

Atomic weight. We bring that term out from storage, dust it off, recall having heard it somewhere. Chemistry! It is back there with words like *mole* and *chemical formula*, words we would rather forget, the intellectual equivalents of spinach and broccoli. I once dreamed of atomic weights: myself on one side of a huge pan balance, metallic

*The astronomer Sir Arthur Eddington begins his book *The Nature of the Physical World* with a fascinating discussion of the "two tables." Which of Eddington's tables is "real"—the table described by modern physics, or the table of our everyday sense experience—has been the subject of considerable philosophical debate ever since.

cobalt being shoveled from a pickup truck onto the other. All I had to do was count the atoms and I would have the atomic weight of cobalt. I woke up sweating, wondering where I was.

Dalton was far more elegant. While his method could not get you the actual weight of an atom, it could get you its *relative* weight. It worked like this: Suppose you had some water, say eighteen grams of water, enough to line the bottom of a coffee cup to a few inches' depth. If you could break that water into its simplest parts—its elements, hydrogen and oxygen—then you could weigh those parts and find out just how much hydrogen combined with how much oxygen to give water. When this was done (and in practice it was done by passing a current through the water, collecting the gaseous hydrogen and oxygen at the electrodes), it always gave the same result: sixteen grams of oxygen at one electrode, two grams of hydrogen at the other. The ratio of the "combining weights" was 16/2 or 8/1.

Did that mean an oxygen atom was eight times heavier than a hydrogen atom? It all depended. It depended on what you thought the formula for water was. If, like Dalton, you thought that water was represented by the formula HO, one atom of hydrogen and one of oxygen, and that you had, therefore, collected equal numbers of oxygen and hydrogen "particles" at your electrodes, it made sense to say, from the combining weights, that an oxygen atom was eight times heavier than a hydrogen atom. But Dalton's "water" would have made a strange, strange world. Not our world.

Suppose instead water was made. up of two hydrogen atoms for every atom of oxygen, the way that Amadeo Avogadro said it was, in the early nineteenth century. Then what would happen? Now there would be twice as many particles of hydrogen as oxygen, but still the same weights, still sixteen grams ·of oxygen to two grams of hydrogen. If water is really "H_2O," then oxygen must be *sixteen* times heavier than hydrogen.

The trick was to know combining weights *and* formulas so that you could get relative weights. Then you could decide that an oxygen atom weighed sixteen times as much as a hydrogen atom and that carbon weighed twelve times as much. This simple trick, this weighing of atoms, is why we remember John Dalton as the Isaac Newton of chemistry.

Still, those were only *relative* weights, and it took another fifty-eight years, the congress at Karlsruhe, and the eloquence of Cannizzaro to work out a consistent set of them. It took another fifty years to get *absolute* weights—how much an atom really weighs in comparison, say, to an autumn leaf. This could only be achieved once Planck and Einstein had determined how many atoms there were in twelve grams of carbon, or one gram of atomic hydrogen, or sixteen grams of atomic oxygen. It turned out to be a lot, and it turned out to be the same in every case: 6.02×10^{23}. Avogadro's number. A mole of particles.

Alas, more broccoli. Yet for the chemist these were the good old days, the days when things were simple, when the atom made sense in a good intuitive way, and when the world it rested on did too. All that changed with Rutherford, and in a way it changed with a single experiment.

By the early part of this century, people knew that Dalton had not quite gotten it right. The atom was a thing of parts, charged parts in fact, fleeting bits and pieces of electricity. J. J. Thomson and Robert Millikan had weighed the electron, defined its charge. Thomson had even proposed a model for how electrons might exist in atoms. What he envisioned was an English pudding, a pudding of positive charge in which were embedded plums of negative electricity—the electrons. The "plum pudding model," however, mouthwatering as it was to contemplate, was put to the test by Rutherford and found to be wanting.

In 1909, at Cambridge University, Rutherford took a very thin sheet of gold foil and subjected it to a barrage of charged particles. He expected them to rush through the foil, to be bent only slightly from their course. And most of the time that was exactly what happened.

But on occasion, something strange occurred. Some of the particles, moving at half the speed of light, actually rebounded, came back and hit the source, like a tennis ball ricocheting from a wall. Rutherford was stunned. In a letter to a friend he said:

> It was quite the most incredible event that has ever happened to me in my life. It was almost as incredible as if you fired a fifteen-inch shell at a piece of tissue paper and it came back and hit you.

What was causing this flimsy sheet of "tissue paper" to withstand the mighty rush of an "artillery shell," and more, to turn it around and send it back from where it had come? Rutherford thought he knew, and his calculations bore him out. The particles could have been repelled with such force, he reasoned, only if all of the positive charge in the gold atom were compressed into a hard, dense, fiercely compacted point of matter: into a "nucleus." But where were the electrons? Rutherford argued, and his brilliant student Niels Bohr showed mathematically, that the electrons must be orbiting this nuclear sun like so many tiny planets, like vacant moons. Suddenly it was no longer possible to think of atoms as Daltonian billiard balls, or even as plum puddings. With this single experiment, the atom, upon which the hard back of the physical world rested, became mostly emptiness.

Imagine the rounded dome of a great cathedral on a quiet morning. In its center, on an updraft, is supported a single dust mote, a particle of mica. Between this speck and the great dome there is nothing. Place yourself there on that ceiling and look across the broad enclosure, rose-colored at this hour, toward that glittering point. You are at the distance of the first electron shell, looking inward to the heart of matter itself, whirling and caught in the force of that insignificant nub, the nucleus. And you can barely discern it, barely make it out. What you see, the great fact of your experience, is space pure and simple, empty space, awash over matter like a sea.

That is what solidity—the solidity of tables and chairs, of rocks and mountains and ice—became after Rutherford. An illusion. This table in the garden is no more a plenum than was the latticework of lights that had lain below us in the Hawaiian night. Rather it is a system of shimmering forces, as nuclei rein in the restive electrons, and electrons occupy space the way a turning fan blade does, not by filling it, but by making it impossible for anything else to fill it.

▼ ▼ ▼

What once seemed full now seems empty. What once seemed so durable now seems evanescent; what once seemed permanent now seems an accidental gathering of particles—gulls on the beach bound for dispersion. For at this level of analysis, this table, matter itself, has too many points of entry, too many microscopic toeholds, too many interstices, too much motion. It can be teased apart, the way manganese is teased by water from basalt; disassembled the way rock is disassembled by wind, layer by layer, atom by atom; the way iron oxides are dissolved and dispersed in the depths of a summer-stratified lake. This table can be turned to smoke, recycled back to carbon dioxide, sent on its way through the wild troposphere, if only enough heat is applied—a small flame will do. Against time and chance this genial gathering of molecules is nothing.

So I lean against the garden table, in the shade of the monkey puzzle tree. As furniture this bit of nothing is entirely adequate. Though now, in the wake of these reflections, it is less familiar.

▼ ▼ ▼

As it turned out, the physics of Rutherford was only the beginning. Bohr had then to see the possible arrangements of the electron in hydrogen, its planetary paths around the nucleus, its circus leaps and arabesques through the void, the lines of visible light that it cast as a spectrum from eerie tubes in darkened rooms—red, green, blue-green, violet (these and no others). Matter and light linked; the hues of the Swiss schoolteacher Johann Balmer, caught in the net of numbers. Heisenberg, Born, Schroedinger, Dirac, Jordan—the quantum theorists of the "golden age"—had yet to capture the electron as wave, as barn swallow—here, there, everywhere, mere mist of charge. They had yet to give deepened logic to the elements, to the play and recurrence of their themes: the nobility of platinum and gold, the vigor of metallic potassium and cesium, the utter vivacity of fluorine and chlorine.

After two short years, from 1925 to 1927, you could at last understand why hydrogen behaved as it did, its vehement attachments, its liaisons with nearly everything. And the same with oxygen: Now you could rationalize its valence, its self-involvement as a diatomic gas, its undisguised propensity for metals. The red and orange rusts of iron followed exactly from this. And Mendeleyev's elements, stacked in neat empirical columns, strung in long, sonorous rows, finally sounded, as Mendeleyev had known all along, like a lovely sonata.

So there is a trembling at the heart of matter, at its very core a disquiet. Things oscillate and turn, twist to breaking, collide and sunder, re-form anew. Generation comes from this. And decay. The way a lake rolls and stratifies and rolls again. All that we see is change. The shadow moving on the mountain is the Earth's moving, the sun's moving, the explosion of hydrogen against hydrogen far away becoming helium, casting light, spitting photons through vacuums of folded space. Light lands on the mountain, warms the stone, sets it to moving faster deep within itself. You can measure these things, you can feel them against your skin, recon their beginnings as events, as births and deaths by the billion. Warming yourself on a simple rock, you know catastrophe upon catastrophe and you are gladdened by it all, knowing this is the way it has to be, it can be no other. We were given only one world and this is it: reckless, dangerous, sweet, forever in flux. The sun's death raises the brook in the hills, warms the stone, warms the skin, warms the very soul along its way.

This trembling cannot be stilled, not down at the farthest zero, not on the still life of canvas. Cézanne's apples spilling on the table, spilling on the cloth. What is still? Not the apples Cézanne faced, brush in hand, waxed skin glowing red, yellow, green, like a spectrum. Not the painting, not now, not ever, not in reproduction. In the wine there is change, in the straw, in the cloth, in the simple fruit. You cannot hold on to it, it is always passing, always gone. Cézanne knew this.

This change has been written into things like a law: electrons on the wing, atoms shaking at their very core. Things build and fall apart. Only the pattern remains, only the name. The mountain is rising, is being worn down, never the same mountain. Look twice, you will not see what you saw before. The flanks, the peaks, the ridges. Even the pattern changes: this is the vision, and physics sings of it. And chemistry. And geology. And biology. Thus, from Rutherford's physics and what came in its wake the great cycles of the world emerge as necessities. The vision of Ecclesiastes finds its theoretical underpinnings. *Beauty and death are everywhere, and everywhere are entwined.*

I think it was Varner who reminded me that this journey into the heart of matter, which scientists always associate with the early decades of the century, coincides neatly with the quest for the Pole. Scott's Discovery Expedition, which came through these very gardens (1901-4), extended the human presence south to 82 degrees latitude during the very years that Planck was probing the discontinuity of physical process and Thomson was sketching the contents of the atom. These were the years of Soddy's work on radioactivity, of Lenard's on the photoelectric effect, and of Roentgen's on X rays. By the time of Scott's Terra Nova Expedition, Einstein had explained the photoelectric effect, had shown that Brownian motion was the outcome of atomic and molecular events, and had published his epochal papers on special relativity. Bohr was only a year away from his quantum trilogy on the constitution of atoms and molecules.

There was a probing of limits in that age, an effacement of distance, a turning outward and in at the same time that shadowed forth a deeper need. This intellectual passion, this desire to take the measure of the world, to map the territory in which we dwell, shone so brightly that it all but masked the more common motives that were surely there as well. The pure ebullience that seemed to flow out of Thomson's laboratory, the playfulness of Bohr, with his ping-pong and his animated walks through the Tivoli, the swagger of Rutherford in Manchester claiming of the electron, "why I can see the little beggars there in front of me as plainly as I can see that spoon," all suggest a grand adventure under way, motivated by a simple need to know.

▾ ▾ ▾

The warehouse stood in the Antarctic facility near the Christchurch airport. It was a cheerless, functional place. Cement floors, dark walls, cold even in spring, unless the wide doors were thrown open. Before this morning there had been a sense that the journey to the Ice was not yet real; that I would roll over soon, wake up, stare out the window of my own house into the gray Ohio sky, into the leafless redbuds, into my own backyard. But here, now, it became real.

There was a seascape on the wall, a black and white photograph, eight feet by ten, taken by William Curtsinger. In the photograph, a freezing southern ocean heaved, under gray skies, the bitter cold sea blending into low cloud, sky and water dissolving in gray. No ice floes, no pack ice, no research vessels. Only the sea itself. I knew the photograph must have been taken twenty years ago, when Curtsinger worked these latitudes. But it held an instant in time that would play itself over and over again; that at this very moment, as we dressed, was being relived. In this photograph, the southern ocean seemed to brook no intrusion. "Go home to your cities," said the grim sea. "Go home." And I once heard of a scientist who, having gazed into this scene for several minutes, thought better of the journey south, handed his clothing back, left the warehouse, and returned to Los Angeles.

Eventually we approached our plane. I climbed through the door into the hold, felt my way along the webbing and canvas seats toward the back. Pipes and wires ran the length of the fuselage, and huge crates, like some dark cordillera, tapered into the tail. I groped my way into a slot between two Navy men and sank into my seat. The plane shook. I took the yellow ear plugs, rolled them into points, and began to insert them into my ears. Mike was yelling at the top of his voice, but I could hardly hear him. "We're finally going," he shouted. "Finally. I want to get to the lakes." I nodded and pushed the plugs in.

Moments later we were airborne, above clouds, the tiny windows silver with the morning light.

▼ ▼ ▼

Six hours out of Christchurch, I walk toward the back of the Herc. I can see my breath. The wool shirt, once close and prickly and hot, is now warm and comfortable. I lean against the cold skin of the aircraft, slouch toward the window, and peer out.

The ice below me is spreading slowly, serenely, poleward. It is only this water molecule: these simple threads of force thrown and repeated over and over, a pattern of matter foreordained by the way that ten electrons weave themselves among three nuclei. The hydrogen of this molecule gropes for the outreaching electrons of that molecule, and the hydrogen of that molecule seeks its opposite charge on the next. If the temperature is low enough and the gropings are not unrequited, ice will form. Project these patterns over thousands of miles, thicken them, give them the hard, illumined edges of sea cliffs; let them gently curve with the Earth's curve, be unreceptive of light, and you have what is below me in clear view, the glowing white underside of our globe. Greater Antarctica.

Varner is up now. He slowly works his way toward the back, squeezes by a wall of wooden crates. He hands me a crumpled sheet of paper as he passes. On it is a list of flight times. "CINCI to LAX, 4 hours; LAX to HONO, 5 hours; HONO to AUCK, 10 hours; AUCK to CHCH, 1 hour; CHCH to MCM, 8 hours." "Only two hours more to go," the sheet reads. "I want to be on the ground. Or the ice. Or whatever it is that's down there."

Soon everyone is awake, sitting rigid, hands folded in laps, knotted against the slick fabric of the windpants, the hoods of parkas pulled snugly around faces. We are banking toward Black Island, then away, coming in over McMurdo Sound, making toward the ice runway. The landing is smooth and there is a mild sense of deceleration as the wide skis of the aircraft touch down. The Herc taxies and in five minutes comes to rest; the engines fall silent, the door swings open. The gloom of the fuselage is broken, suddenly shattered by an almost supernal light. We have arrived.

We file through the narrow door. Outside there is silence, except for the wind and the crunch of boots on dry snow. Desert snow. I stagger a little at first, awkward under the weight of my flight bag and thirty pounds of clothing. I squint, trying to adjust my eyes to the glare.

Off to the right I can see Ross Island, and Mount Erebus, a line of steam at a perfect right angle to the cone. There is a cluster of huts to the left, like photographs of old Dakota mining camps. And then this vast whiteness in front of me and behind, and, far away, mountains running off behind mountains behind other mountains into the distance. A continent of mountains lost under the long ages of snow. Thirty million years of snow. I think of my father, what he said about openness and light. How these set you free. I look at Varner, who is turning slowly in circles. I put a hand on his shoulder. He says only, "This isn't real." Then he repeats it.

FOUR

The Map

I was remembering my father's maps, those old Texaco numbers, creased into falling lint at the folds, folding open with the smell of paper dust and raw heat from the glove compartment. I was thinking of him opening his maps on the front seat of the Pontiac Chieftain, moving the wet, well-chewed El Producto cigar around in his mouth, exhaling a blue cloud of vapor that hung and hung until it got stale in the hot car. And we were in the back, eating these red-stained warm tomato sandwiches he had packed, wondering what curvy switchback of a road he was going to head down next and whether we were going to make it through the afternoon with our stomachs intact. And his head was down now, the map crawling up my mother's arm, and his finger was tracing along some thin spaghetti of highway that ran near battlefields and the quilt of farms and over the river with its ferry and down to the sea.

And I wanted us just to stop for the day, get out, walk over behind the small white motel units beyond the neon signs into the fields and throw baseballs, make them disappear into the red evening sky and then reappear as plunging arrows onto the green earth. But he had his map now, shredded in the creases, red- and blue-veined and smelling of oil and heat, and the dead gray ash of his cigar was falling onto it, and he was brushing the dry ash away.

My mother was wearing a white dress or maybe it was a blue dress with a white collar, and she was saying nothing, just looking off into the distance beyond the chrome war bonnet on the hood. The map was my father's instrument of navigation on the long eastward journey through the strawberry and melon and corn roads of Pennsylvania and New Jersey, and through the streets of the seaside city with its wooden three-story guest houses and its seaside restaurants and its piers jutting into the Atlantic. With his map he would find us a room and a meal and a bed to sleep on. Most of all, with his map he would find us the churning, magnificent sea. The sea filled with light.

▼ ▼ ▼

NIGHT IN ANTARCTICA is so little different from day. At McMurdo you sense only a slowing of the pace, a diminishing of machine noise. The orange helicopters, with their rigid schedules, sit motionless on the pad, their blades sacked and roped into place against the wind. The sun at this time of year merely circles the horizon, a pale disk low in the sky. In full daylight, the streets are mostly empty. People have gone inside, into the warmth and familiarity of the television room with its sofas and magazines and beer dispensers dropping Coors and Michelob for half a dollar into your hands; and into the officers' club, with its antiquated popcorn machine, still there, still off in the corner across from the bar; and into countless dens and warrens of ping-pong and eight-ball; and into solitary rooms with their books and pens and papers edging away the distance and loneliness. In my own room, I pulled down the window shade, closed the drapes, and pinned to the rough fabric a map of the continent. I poured some bourbon into a plastic cup, pulled up a chair, propped my feet against the wall, and began to study the white and blue shapes that hung before me.

A map can do strange things. On a flat surface the incomprehensible world lies stretched and somehow intelligible, its myriad particulars brought together in a single, unified whole. Out of the chaos of coordinates and compass readings and place names, an order emerges, a system of relationships that seems somehow complete and enduring. With a few pins, or with a few words of historical commentary, the dimension of time can be added, the progression of events can be traced. In a single glance an entire universe of content and association can be reconstructed in a thousand different ways, each remembrance colored and shadowed by whatever unpredictable gloss the mind lays over it.

I remember once, while reading *Moby Dick*, I mounted a map on a large slab of cardboard over my desk and festooned it with pins that eventually traced the journey and fate of the *Pequod*. In greens and reds and blues, the pins marched across the breadth of the oceans from the tiny hook of Cape Cod, eastward to the Azores and south down the Mid-Atlantic Ridge, curved westward along the coast of South America to where the Rio de la Plata empties into the sea. Then they moved in a straight line northwest to St. Helena, south around the Cape of Good Hope into the Indian Ocean, northeast into the South China Sea, and off toward Japan. Off Fanning Island, near Samoa and

Tonga, I stuck a black pin into the map, where the sea had rolled over the ship's mainmast, leaving not a trace of its passage. The map that hung before me was like this: a symbol of movement and passage and quest, of objects vast and minuscule creaking and ambling and probing through time. A symbol of order, of things understood, connected and whole.

The click of metal, the sturdy door opening, brought me back. Varner had returned from the lab. There was a trace of isobutanol from the phosphorus analysis still clinging to his clothing. He sat down at one of the desk chairs and said nothing, as though the presence of the map in the darkened room was something that needed no explanation. After a few minutes he turned to me and said, "You know, this place is starting to scare me."

He was leafing through the unmarked pages of the journal he had bought in New Zealand. "I don't think I can write about this," he said. "It's just too immense. Too damned hostile looking." He was pointing up to the map, inscribing a big circle in the air. "There is no language for this."

He moved to his bunk and lay down, his face over against the wall. "I'll tell you what's out there," he said. "Nothing. Nothing that I can understand. Nothing that you can understand either, even though you pretend." He fell silent for a minute. Then he turned over and rubbed his eyes. "Don't say anything to Mike and Walt and the others. They wouldn't understand. I'm not sure I do."

Varner turned toward the wall again. His plaid shirt picked up the light filtering through the map that covered the window. "Do you know what light does to our body's sleep cycle?" he asked rhetorically. "Do you know how unnatural twenty-four hours of daylight is?" He was speaking in a very low voice, so that I could barely hear him. "We're meant for light and dark. Not for this. We're not meant to be here at all."

I didn't know what to say, so I let it go. He'll get over it, I thought. It's early. Things will change. He'll find the words he needs. I let my eyes drift back to the map. I became more interested in thinking about the continent in other ways, about how it had come to be here, so isolated and ice-filled, so remote from our history, almost invisible. It was immense, as Varner had said, almost too immense to have drifted here into these latitudes, undiscovered and nameless until so recently. Until just last year, you might say.

I ran my finger in a clockwise circle, whispering the names that appeared over the background of the ice: Queen Maud Land; Wilkes Land; Marie Byrd Land; Ellsworth Land; Palmer and Graham Lands, curving off toward Argentina. Outward from the Pole, I traced my finger over the coasts: Princess Martha Coast; Princess Astrid Coast; Princess Ragnhild Coast; Ingrid Christensen Coast; Queen Mary Coast. And over the ice shelves: Ross, Ronne, Filchner. And the capes: Darnley, Elliott, Goodenough, Adare. And the seas: Weddell, Ross, Amundsen, Bellinghausen. Name after name, kings and queens and princesses, explorers and statesmen, ordinary seamen and scientists—I spoke them aloud, as my finger traced the disk of continent, traced its 5.5 million square miles (as large as the continental United States and Mexico combined!), its circling geography of plateaus and ice-buried mountains, its indented coastlines, like cockles and scallops, until I felt dizzied by it all. This naming, what did it mean? Could you possess this? And once you did, what would you have? A handful of water. Better to bless yourself.

The shape on the bed rolled over. "You forgot Varner Land and Cape Varner and the Varner Massif," it said. I laughed, took out a pen. Stood up and moved the few feet over to the map. I drew a line through the word *Ross* and above it wrote *Varner*. The Varner Sea. What did it matter? Varner smiled, rolled back to the wall, trying to sleep.

I sipped my bourbon and continued to scan the map. The Dry Valleys of Victoria Land, where we would soon set up our field camps, lay across the narrowing wedge of McMurdo Sound. They appeared as little patches of earth pressed between the sea and the ice shelf. On this scale they were nothing more than a slight discoloration, a brown edging woven into the linen whiteness of the continent's ice. In fact, the valleys were part of a great mountain range, the Transantarctics, that split the continent into East Antarctica and West Antarctica, into provinces of rock and ice-covered archipelago. Like the Mid-Atlantic Ridge, which lay beneath the *Pequod*'s hull as it moved midwater between the rifting continents of Africa and South America, or like the ragged earthseam of the East Pacific Rise, the Transantarctics are largely submerged—submerged not by water but by a mile or more of ice. One mile of ice!

On this map, just beyond McMurdo Sound, some words are

printed in the vicinity of the McKay Glacier, near a patch of brown exposed rock:

Antarctica once supported vegetation. Anthracite coal was found near here in 1961, along with petrified wood; one log measured 17 feet long and 15 inches in diameter.

I sat in the wooden chair at the desk near the window and propped my feet on the scarred wood. I thought of the Carboniferous Period, lush and seething and filled with life like the Amazon jungle, steaming and vaporous and crawling, a land of softness and sudden death. And it was all out there, just beyond where the low sun was turning the map transparent, just beyond the ice. Three hundred million years ago, when none of this bitter cold would have seemed possible, there was a riot of life. The rich stores of atmospheric carbon dioxide, themselves partly responsible for the planet's ancient warmth, were being fixed by green plants, being turned, in yet another alchemical display, into cellulose and lignin chains, into hydrogen-bonded cords of DNA twisting and disappearing into thin air; into the mat and fiber of giant ferns. And in the watery warm abundance of those ages, the Carboniferous ferns and calamitids choked the swampland in sheer mass and number of greenery, decomposed in the sinking mallow, and were buried under a wash of sand and silt and clay. All of this stored carbon, banked in the darkness of swamp, carefully hidden away from molecular oxygen, evolved slowly into coal. The carbon dioxide of that distant atmosphere first became the stuff of plants, and then, in the long transformations of carbon through the agencies of bacteria, heat, and pressure, the stuff of coal. And it lay out there now, among the exposed rocks of Victoria Land, testimony to all that was once lush and vibrant and breathing—and past.

South along the Shackleton and Dufek coasts, then southeast among the Queen Maud Mountains, my eyes traced to Graphite Peak. On the map, it appeared below a sweeping blue arrow and several lines of text that describe what happened there. It was near Graphite Peak, in January 1968, that Peter Barrett came upon a fossilized bone, a jaw fragment lodged in the coarse Beacon Sandstone that forms the mountain. When it was sent to Ohio State, the fossil was identified as an amphibian land vertebrate that had lived some 200 million years ago—Jurassic time.

The following year, in November, a much richer store of fossils was discovered at nearby Coalsack Bluff. Among the remains that lay in the bedded sandstones were those of Lystrosaurus, a low-slung, powerfully built creature, sometimes referred to as a therapsid, or mammal-like reptile. In reconstruction, Lystrosaurus resembles a tapering barrel set on four short, muscular legs. Two short tusks and a pair of huge, sad, high-set eyes characterize the skull.

To imagine the world in which it lived is to imagine Antarctica as a different place. The Antarctica of Carboniferous, Permian, Triassic, and Jurassic times was everything that the continent today is not: a land of warmth and wetness joined with other lands in some great Earth-wide conspiracy of organic wealth. To geologists and paleontologists, the sandstones and fossils at Coalsack Bluff conjure a diorama of lakes and rivers and braided streams where, under the fronds of giant ferns, Lystrosaurus roamed and grazed, plunged in mud like a small hippo.

On the map that hung before me there were stories of coal and reptiles and therapsids, things threading their way up out of the past. The continent had been a garden, in the way we think of gardens, wild and serpent-filled and warm. And it had not always been here, in these regions of isolation. It had creaked and broken and split and jerked mountains straight into the air and had torn itself asunder, stretched its rock like taffy, spinning away, out of the reaches of South America, trailing islands and peninsulas like bits of cosmic dust. The story the geologists have written, from bone and rock and ripple marks, from ancient rain prints recorded in stone, is long and loose in its details and so recent in its writing that it can almost be called contemporary fiction, were it not surely science.

In the late Permian, the surface of Earth was arranged differently. The continents were welded into the great landmass of Pangaea, the land that was all lands. To the north lay Siberia, North America, Europe, and Asia. To the south, below where the Sea of Tethys cut inland at the equator, lay the connected lands of Gondwana: South America, Africa, India, Australia, and Antarctica.

Scattered across the old Gondwana are the rock-bound clues of this ancient geography. Glossopteris, the giant fern whose imprint can be found in the rocks near Graphite Peak and Coalsack Bluff and throughout the nearshore mountains of Victoria Land, has been discovered in broad zones that sweep across South America, Africa, India, and Australia. The squat Lystrosaurus appears also in the fossil records of Africa, India, and Australia. These and other fossils document beyond reasonable doubt the association of land to which Antarctica once belonged.

During the nearly 100 million years of Triassic and Jurassic time that followed, Pangaea began to rupture, first into Laurasia and Gondwana and then into ever smaller units. Driven by the currents of the Earth's mantle, the continents scattered slowly, centimeter by centimeter, until they assembled in more equable arrangements across the face of the Earth. By late Jurassic, some 150 million years before the present, Africa and South America had separated from the landmasses that included India, Antarctica, and Australia.

Later, India would break away and then, 90 million years ago, Australia. Eventually, Antarctica was sent to wander in these latitudes of cold and brilliant light. Once it had centered itself here, had begun its counterclockwise drift about the south geographical pole, its interior became a platform for the winter-long settling of snow. The cold, circumpolar current that developed soon after the parting of Australia further isolated the land from the warmth of tropical seas to the north.

Perhaps the ice began to build here 35 or 40 million years ago in what is called the Oligocene. What snow had accumulated over the months of winter darkness no longer disappeared in the light of summer. Gradually, beginning inland, in the cold heart of the interior, or perhaps high in the mountain passes, ice sheets and glaciers began to form. Whatever plants and animals had hinged their fates to this cooling land were now doomed to seek refuge within its confines, for the last connections to the north had been severed. As the ice sheets expanded, as the dark winters became unendurable, life would have been driven to a few safe havens near the coast. Finally, only those creatures that could find refuge from the killing wind and cold, that could withstand the winter night, survived.

We think of it as nothing unusual, this white draping of ice across the Earth's southern Pole. But that is because it has been with us from our earliest childhood memories—the map above the chalk ledge, the lighted globe in the corner, the geography book with its torn brown cover—as a simple whiteness, as a presence about which we are expected to know, and really care to know, nothing else. Only that. The whiteness of Antarctica, in our conceptual geography, is its right and proper attribute, its only attribute. Unlike Africa or Europe or the islands of the Pacific, which crackle with associations, Antarctica brings forth nothing. That there is something anomalous here in that symbolic whiteness, something calling out for explanation—for every sixth-grade teacher to stand on the desk and proclaim, "Look! Behold this utter strangeness that I am pointing to, this ice cap, this continent!"—never occurs to us. At home, the bottom of my globe is covered by a plate that obliterates everything south of 75 degrees latitude. On the plate it says, "replace with 15-watt, 120-volt lamp only."

We think of it as nothing unusual, and yet this polar whiteness, this ice cap, is one of the rarest and most spectacular of the Earth's features. From space, astronauts who have witnessed it say that it shines like a great, blinding light, a rival moon in the night sky, the most radiant of lands, returning the sun's light, magnified in beauty, a thousand times. Only twice in the past 570 million years, since the dawn of the Cambrian and the explosion of life, has the Earth been so configured in its continents and so composed in its atmospheric chemistry that it could sustain the glory of a southern ice cap. At least two conditions had to be met: there first had to be a landmass spanning the Pole, a continent on which snow could begin to accumulate and build; and the temperature at the Pole had to be suitably low, to prevent the summer undoing of what winter had brought.

During these rare geologic times, when the Earth tended more to the inorganic than to the profusion of life, a chill swept over the planet. It was a chill that we have come to know in our own insignificant slice of time, a chill whose source we trace down the deep curve of the Earth to the ice slurry of spinning seas and finally to the cold well of the continent itself. Antarctica, lying so roundly across the Pole, alters the whole of the Earth's heat balance. For under more normal geophysical conditions, light from the sun would be absorbed by seas or continents, reradiated at longer wavelengths, and absorbed by the gases of the atmosphere. But much of it is now reflected by the brilliant ice and scattered back into space. Thus, as the ice cover of the continent built gradually outward from its center, its ability to act as a global reflector became enhanced: the growing ice sheet cooled the Earth and provided more suitable conditions for its continued growth.

If the ice cap functions as a kind of sink in an enormous heat engine operating between the equator and the Pole, it also acts as the silent engine of ocean circulation. The dark of Antarctic winter is the time of transformation, of state changes, of water becoming ice. In infrared satellite photographs that record the reaches of the ice in shades of pinks and purples, you can see an almost seasonal breathing of the Southern Ocean. In September, near the southern winter's end, the sea ice appears in full extension around the scalloped edges of the continent. By February the satellite image shows the ice contracted to the continent's margin, the swirled pastels drawn into tight little harbors in the Ross and Weddell Seas. In winter, at its maximum, the sea ice covers 20 million square kilometers, an area three times the size of the United States and nearly five times the ice's summer extension.

When ice freezes, it excludes most of the sea's ions: sodium, chloride, calcium, bicarbonate, magnesium, sulfate, potassium—all are sent back into the slurry of the sea. As the ice sheet builds outward from the cold core of the continent, the cast-off brine streams downward into the ocean depths, forming, over vast areas, what oceanographers call Antarctic Bottom Water. The sinking of this cold, oxygen-rich water in the frigid gloom of the austral night is the piston stroke of deep ocean circulation, of the sea's great exchanges between atmosphere and ocean floor. In times when the ice cap melts, the ocean basins become, in the constant downward rain of organic matter, stagnant, like the hypolimnia of summer-stratified lakes. When the ice cap melts, the dynamism of the surface Earth diminishes; the power of its winds and waters is quelled.

It seems surprising at first, especially to those of us accustomed to thinking of the vast lakes of Africa and Asia and North America, that so much of the Earth's water should be locked in the immobility of Antarctic ice. But it is true. If all of the planet's lakes—the Victorias and Superiors and Baikals, the Great Bear and Slave lakes of Canada, the Caspian and Aral seas, and the dazzling lakes of the Qinghai Tibet Plateau—were grouped together and frozen in place, they would form but an insignificant fraction of this, a blue eye in the vast corner of Queen Maud Land. If the Antarctic ice cap were suddenly to melt, sea level would rise sixty meters. Every island and fishing village, every port and coastal plain—the Fijis and New Bedfords, the New Yorks and Stockholms, the low-lying Louisianas of the world—would be inundated in this summary redrawing of the earth's geography.

If ice is the dominant feature of the land, then here, in this sector, the name "Ross" is surely the dominant feature of its map. Born in 1800, Sir James Clark Ross was to become one of the most accomplished of the nineteenth-century polar explorers. By 1839, when he was dispatched to these waters by the British Admiralty, Ross had already established an enviable record of seamanship and command in the Arctic, beginning with his discovery of the North Magnetic Pole eight years earlier. In 1841, Ross's ships crossed the Antarctic Circle. With wind-filled sails and in the full radiance of Antarctic summer, the three-masted vessels pressed among the pack ice, through bergs of refractive splendor. On January 11, Ross became the first person to lay eyes on the coastal mountains, those landforms so unexpectedly arisen from the sea, with their illusory peaks and valleys.

His very lands lay just beyond the parchment-colored curtains of the room, beyond the map pinned to them: the Ross Sea, the Ross Ice Shelf, Victoria Land, with its untouched lakes—Bonney and Vanda, Miers and Fryxell and Hoare—and finally Ross Island itself. From this tiny outcrop of volcanism, this island of flame in a sea of ice, the threads of journeys lie in tangled skeins upon the map. Scott's route to the Pole leads from here, across McMurdo Sound over the ice shelf, south along the Shackleton Coast, then up the Beardmore onto the Plateau. On the map it is a line of black crosses marking elevations to that horizonless, windswept desolation.

It was January 11, 1841. Queen Victoria was a young girl. Charles Lyell, the great uniformitarian, was on the verge of transforming geology into the science of slow accretions and wastings, the science that would eventually rescript the world to accommodate its mountains, its seafloors, its ice caps the size of moons, and its time—its bottomless wells of time. A week earlier, in America, Melville had set sail from the Massachusetts harbor of Fairhaven on the whaler *Acushnet*, bound for Cape Horn and the Pacific. But in these latitudes, Ross had just entered the Land of Light, had just named the chimera of mountains risen from the sea after the young girl. He had called it Victoria Land. After millions of years of isolation, the name actually mattered.

FIVE

Science and the Shell

There is an image from my childhood, stored in the soft gatherings of large mole-cules, warm and safe and happily imprecise, as memories are, unlocatable on any grid of coordinates and yet so easily accessible to me that a mere word or smell or touch can call it forward out of its coiled sleep. It is an image of the town on a win-ter night, the still air crowded with the downward drift of snowflakes settling into the dark crooks and branches of the trees, building on the roofs of houses, up to the chimneys with their billowing smoke. I have been running through the buried streets, the bricks long since disappeared, running and hiding behind the Chevys and Pontiacs with their chrome grills, crouching and taking snowball-aim at Roger and Janet and Don, then crouching again as the loosely packed snow explodes with a whoosh! *from the car hood in a shower of crystal under the creamy glow of a street lamp. But it is quiet now. They have all gone home to the warmth of living-room fires and to the flicker and glow and magic of black and white television; they have left the neighborhood, the world, the silence, to me.*

From the hill I have climbed, I can barely see the spires and domes of the Gothic and Romanesque churches, and the beacon of the Gulf Building signaling storm; and the mill stacks reddened with fire, tracing the river south. Shawls and linens fall across the land, soften its shapes and edges, turn it undulant and com-mon-hued. A universe without boundary, without the bold line of curbstone and horizon, rolls away in hollows and rises and faint shadow, and then curves in upon itself. I fix a single snowflake with my eyes, guide it through the crowded air, let it settle on my tongue like a host. There is a small exchange of warmth as it melts, as the sharp dagger edges crimp inward toward roundness, toward the shape that structureless freedom demands, becomes a droplet of water, then a memory, some-thing to be carried around forever, a ghost lurking among the molecules.

I did not know it then, but on that night, the night of "the Great Snow" as my mother would call it, there came a longing for places I could not name, places I could not position on any map. I became homesick for somewhere I had never been, for someplace that might not even have existed. Somehow, in the strange illumination that lamplight and snowfall made, standing safe and starless under the dusting clouds, without the slightest hint of knowing it, I felt the first gentle pull. I went to bed happy that night, in the high attic room, under the insulated eaves, thinking that I could hear the snow, flake by flake, as it fell on the steep-angled roof, thinking I was walking, as children do, beyond the planet's edge.

▼ ▼ ▼

SILVER LIGHT ENTERED THE WINDOW from the south, fell across the desks and bookcases and gathered in the folds of sheets where it lay in pools. I had fallen asleep on the cot, still dressed in my windpants and suspenders and cotton long johns, and covered over with my parka. When I awoke, Varner was standing by the window, his elbows pressed against the sill, smoking a cigarette. He was staring off across McMurdo Sound toward the Dry Valleys. "Strange place," he said, when he heard me get up. "I can't stop thinking about it."

He was speaking very slowly, measuring his words. "Even in the lab, I'm thinking about it. When I sleep, I'm dreaming about it. I didn't know it would be like this."

He turned around and faced me. He hadn't shaved since we arrived a week ago. His eyes were red. I had never seen him look like this. "You know," he said, "I think Mike could build the flume if I left. Mike and Tim, I think they could get it right."

For the last few days I had thought that he was thinking about leaving. But this was first time he had said it. Hearing the actual words, seeing his face as he spoke them, shocked me.

"Sure," I said, realizing that it would be a week before the next plane left for Christchurch. "But maybe we should talk about it. Maybe you should have some breakfast first. It's a long way back, you know." He turned to the window and continued to gaze out over the frozen sea. I put on a flannel shirt over my suspenders and slipped on a pair of tennis shoes and told him I would be over in the lab for a few hours. It was only five in the morning, and the mess hall didn't open until six.

I left the room, ran down the carpeted hallway, down the steps to the first floor and out the thick, wind-beaten airlock door that stood between the tropical warmth of the dormitory, the Hotel California as it was called, and the raw weather of the streets outside.

In the Biolab, warm air and the smell of hot chocolate mingled in the foyer. A green leather sofa slumped like soft sculpture against the north wall, facing a large Curtsinger photograph of the research vessel *Hero* in icy seas. I ran hot water into a paper cup, mixed in instant coffee, and headed through the narrow hallways that led into the lab we had been assigned. The room was low-ceilinged and large but unusually crowded with equipment: a gigantic steel autoclave, a bank of refrigerators, a UV-visible spectrophotometer, sinks and pumps and benches and stills and precision balances; three desks were set against the outside wall. All of these familiar objects conspired to delude you into thinking that you were in an ordinary place, in an urban laboratory perhaps, off the circling beltway of a city, or in some industrial park, or on a university campus.

There is solace in this trade of benchtop chemistry if you approach it right, give it the care it deserves. I began my work in the stockroom, in the gloom of the windowless aisles that run between the shelves of flasks and beakers and burettes, that border on nacreous coils of Tygon tubing and wedges of cork and rubber and wintery sheaves of glass. I collected volumetric flasks and pipettes, and beakers that ranged in size from thimbles to small buckets. From a second stockroom, the one where they kept reagents and graph paper and odd little cloth-bound notebooks with faded green covers, I gathered sodium chloride and potassium dichromate and a dozen other salts—white and fuchsia and burnt orange—and brought them into the lab. Some of the reagent bottles bore the names of Ernest Angino and Charles Goldman and Alex Wilson, the legendary pioneers of Antarctic lake research, those geologists and limnologists of the early 1960s whose papers I had nearly memorized, had collected in my folder years ago. I felt myself bumping unexpectedly against the past—at once ancient in its shadowed containers of salts and dyes on dusty shelves, and yet, like everything human on this continent, so absolutely new.

I began by cleaning glassware, by taking a one-liter flask from the bench, holding it gently, not quite certain as yet of my own dexterity —in fact, certain of my utter lack of it, like those countless students who will swear that their very entry into a laboratory is enough to shatter, poison, and explode, to create conditions of unchecked mayhem, exodus, and ruin. But hoping against these outcomes, I took the powdered cleanser from the shelf and carefully poured it down the flask's long neck. Water from the tap turned in tight little ribbons against the inner surface of the glass, then spewed off into the great bulb, where it floated and foamed the cleanser into a rising head. I shook the flask, plugged it with a ground glass stopper, inverted it, shook it again, unplugged it, drained it with swooshes and gurgles into the depths of the sink. I repeated this again and again, this ritual magic, this scientific necessity. Then I rinsed the flask, once in warm water, once in cold, and placed it upside down on a wooden dowel to drain. In time I had an array of glassware, sparkling and dripping and refracting the lab, bending its objects into rounded and rodlike shapes, and all of it was whole and unbroken and gleaming, a collage of the glassblower's art.

The glass was only a prelude. Beyond it lay the precise weighing of salts: the noiseless sliding of plastic doors in the grooves of a Mettler balance, the creation of a small sanctum sanctorum of wind-less calm, where the salt of the Earth could be totted up in grams and moles, strung to the third decimal place—a number you could believe in. And then the making of solutions, the pouring of salt grains and crystals from waxen paper, a flood of crystalline light into a crystalline flask, freshly purged and cool to the touch, and waiting like a gourd to be filled, to be "made to the mark," as the analytical texts always said, made up to the "fiducial mark," the mark of firm belief, of trust, the mark of undoubt. For once the meniscus glided tangent to the mark, lay like a sliver of moonlight caught in the flask's slender neck, you knew something with a kind of finality, something extraordinary: you knew the number of molecules or ions that were present, that were there in that volume, that were enclosed in your hands. You knew the molarity, and for all the world it felt as if you had just counted angels.

The fourteenth edition of *Standard Methods for the Examination of Water and Wastewater* stood alone on a shelf where I had placed it, below the lab window. It was thick and serious in appearance. The spine had been covered with a broad swatch of duct tape to keep it from falling apart. The book was filled with my marginal notes, scrib-bled calculations, underlinings, and checkmarks, and there were mauve reagent stains and acid burns that gave it the look of a crumbling medieval text. But it was in *Standard Methods*, in these nearly twelve hundred humorless pages of weights and measures and operations, that water and salt and crafted glass entered into a single system of knowing, of revelation; it was here that the rune of analysis came to light. I turned to page 202, to the EDTA titrimetric method for the determination of calcium. For it was ultimately to the measurement of calcium that we would have to turn if we were to understand the origin and evolution of the Antarctic lakes.

To dip a spatula into a powdered reagent, to probe its texture and its graininess, to draw it slowly out of its confines, and to watch it lump and plate and roll about in the cavity of a porcelain spoon is to be invited to imagine it in other settings and in other times, to conjure up its possible lives, to cast off the notion that anything, even this inert powder imprisoned on a spatula's tip, could be dull or anything other than shocking in its very being. I weighed out a gram of anhydrous calcium carbonate, weighed it out as *Standard Methods* had specified, to three decimal places, 1.000 g. When I added dilute hydrochloric acid, "a little at a time," the white mass at the bottom of the flask began to effervesce and disappear into the hidden state of things dissolved. In a few more minutes, after adding some of this and pouring some of that and making the volume up to the mark, I had a standard solution of calcium, its molarity known beyond doubt. "This standard solution is equivalent to 1.00 mg $CaCO_3$ per 1.00 ml," declared *Standard Methods*.

▾ ▾ ▾

Go to any mountain stream, newly flooded with spring melt, cutting rock under a blue Colorado sky, and cup your hands in the flow. Gather water in your palms, bring it dripping to your lips. What you taste is not water only, but what water makes when it touches stone— that elixir of Earth's minerals swept up in the molecular tide, the height and breadth of Mendeleyev's table on your tongue. And most prominent among those materials will be not iron or silicon or magnesium or aluminum, as might be thought, but calcium, less abundant in the planet's solid frame, yet more easily teased by water from its rock. And you need not be in Colorado or in the mountains at all, or even under a blue sky. A prairie meander in darkness will do, as will any bayou or backwater or oxbow lake you can find. For in all of these, by all of these, the rock-solid platform on which we stand is being weathered and washed away in the mystery time of atoms and ions, and first among these, always, is calcium. Water works into stone like a plant's root hair, pries and etches and corrodes, beckons what ions will come to come. And calcium, more often than not, does, leaving the rock to its own devices, pitted and honeycombed and collapsing to soil, turning to clay. This is happening every second of every day, on every ridge, in every swale and farm field, this silent, unnoticed release by which the world is undone and remade, undone and remade, in the same breath.

Released from rock—from silicates and limestones, marbles and gypsums and dolomites—calcium is moved seaward, held aloft in the flow by pure agitation, by the energy that randomizes and spreads and disperses. Ion by ion, it is weathered from the continents at a rate of over 500 million tons per year, whole mountains carried in stealth, stolen from beneath our eyes, from the unsuspecting land. What happens to these mountains of calcium, to these stones cast adrift?

At first glance, it seems these mountains go nowhere, that they simply disappear from the Earth—for while the major ions dissolved in the world's rivers are calcium and bicarbonate, mysteriously, the seas into which the rivers flow are not filled predominantly with either. Instead they are of common salt, sodium and chloride.

In 1865 the oceanographer Forschhammer wrote on just this subject, but in words that were so wonderfully prescient and delphic that decades of soundings and samplings and deep dives may yet be needed to give them full meaning. What he said, in that dawn of ocean science, was this: that the chemistry of the seas was only wanly foreshadowed by the chemistry of the rivers; more important were the blind couplings of atoms and ions. Forschhammer knew, somehow knew, long before the *Challenger* slipped its moorings on the Thames in the spring of 1885, that what happened in the sea, that what alchemy the sea performed on the dross of the land, was everything. And that alchemy was most visible, most spectacular, in the transformations of calcium.

▾ ▾ ▾

I remember summers near Cape Cod, after my father died, after the trip he and my mother had taken to Nassau. My mother and my brother and sister, my wife and daughters and my niece would go there, to the old rented house, built in the decade of Forschhammer's paper and set back from the spartina grass behind a New England wall of stone. Those years are all blended in memory now, and what remains is mostly imprecise, a set of time-averaged images: the window of the breeze-filled room that faced onto marshes and, in good weather—on a "five-bridge day," as the natives said—onto islands; and the broad wooden porch with its Adirondack chairs among the delicate pinks of the tea roses; and the great lawn and the harbor clacking with its tall-masted sloops beyond.

My mother's ankles and hands were swollen from poor circulation, even then, so she rarely went beyond the porch. But it seemed hardly to matter. She loved the place, even in her confinement, loved its sea smells and chill salt air and the yard filled with color; and even more, I think, she loved the idea of the sea, its murmuring presence, the way it opened up from inlets and harbors and bays into something deep and mysterious, something she could not comprehend. From that sea, she asked us to bring her a pink shell. It was a request that she tossed out casually, like the good-bye you give to someone heading down the road. "Find me a pink shell today," she would call to us from the porch.

On the rough narrow beaches of the Point, strewn with cobbles and pebbles and glacial debris, our searches were as casual as her request. Over the years we turned up in our wanderings hundreds of shells, brought them to her like treasure, laid them on the wooden floors for sorting. There were thin-shelled scallops, wave-worn and pitted, some nearly diaphanous, their strong ribs converging to a point that seemed far off, infinite, even as you held them in your hand. And keyhole limpets, like tiny white volcanoes, the ribs disappearing down a smokeless caldera. And slipper shells, with their speckled purple and sinuous edges, and their strange epicene nature. And periwinkle and dogwinkle and the spiraling whelk like a fairy-tale castle twisting to the sky. And there were blue mussels and moon shells and conelike augers and razor clams and quahogs, with their growth lines and purple stains. But in all of these, in all of this profusion of form and color, there was not a single pink shell.

I said to her one day, as we were sitting there among shells, "You know, we have the answer to a great problem here."

"And what's that?" she said.

"We know where all the calcium has gone," I replied. "It's all right here. It's on the porch and in the yard. It's in the shells we've been bringing you all these years. We've emptied out the sea of its calcium, and its carbonate too. They go together, you know, those two. All that calcite, with its whorls and ribs and hinges, is right here." She laughed at my geochemical hyperbole and then said: "Yes, but you haven't found me my pink shell yet. And the sea is running out. You'd better hurry."

Even more, it seemed to me as if time were running out, as if things were about to change, with that finality which brings whole worlds to an end. She could barely walk now. Could walk only with one of us holding her, and then only for a few steps, before she had to stop for breath. "Could we stop for a second?" she would say, almost apologetically, "Your old mother isn't as young as she used to be!"

Our last day near the Cape was chilly and overcast. The sky hung heavy over the bay and the yachts held tight at their moorings. From the radio station in Provincetown came forecasts of storm. I hurried up the road, turned through high grasses over wooden skids that lay above shallow water, and headed for the beach. It was early evening, and the wind was beginning to bend and part the spartina like wheat. The waves churned past the offshore boulders and foamed and hissed over the sand. I began to search on this familiar strand, high and low, from the tideline to the marsh, and in the gullies of the small streams flowing to the sea, this time in earnest, for a pink shell. For hours I moved along the sea's edge, moved into darkness and sheeted rain from beyond the islands, moved with my flashlight sweeping the sand, pitching and flinging rocks as though something might be hidden there, burrowed in and buried, something pink.

The land and the sea had become an indiscriminate darkness. With water pouring over my face, I knelt in the sand, gouged my fingers into its unyielding wetness, scooped it outward to my side until I hit something—something rounded and crossed by concentric circles. I pulled the thing up dripping and grit-covered, and then plunged it back into the surf. Under the full beam of the flashlight I held it in the palm of my hand, where it glistened and ran like a sore, a salmon-colored oval, a pink shell. I pulled out my knife and pried it open, took off the top valve, and washed it again in the dark spume. Then I ran up the beach into the marsh grasses, over the flooded walkway and back to the house.

She had gone into the living room, where she sat on the damask sofa, her head lolling against its back, her mouth open in sleep. When I entered the room she awoke, the way she had always done whenever I had come in late. "Bill," she asked, in a voice confused with sleep, "is that you?"

"Look," I said, "I found it. After all this time, I found it." I held out the shell to her. She took it in fingers that arthritis had swollen and bent like roots, and traced its growth lines round and round with her thumbnail. "It's beautiful," she said, in that way that she had of saying things, like a young girl still in the thrall of life and youth. "It's beautiful, just like the one your father gave me in Nassau. How did you ever find it here? They're so rare." Her eyes moistened as she looked down on it, tracing its smoothness, its perfection, remembering something far away.

The next day we left the house and the beaches and the blue harbor and the garden filled with shells, and we never returned. For though I saw my mother many times after that, she was never well enough to travel there. So what had held us together in that house for so many seasons was not to come again. The world had changed.

So the problem of calcium is solved in the building of shells. It is in the coming together of calcium and carbonate, in the celebrated geometries and architectures of the sea. But it is not the familiar near-shore mollusks, nor is it any of the panoply of lightning whelks and sand dollars that we find in abundance on the beaches of the world that accounts for the mountains disappeared from the land. It is the myriad of tiny forams and coccoliths, swarming in the hillocks and valleys of the plunging sea, the open sea beyond the land's last light, that knead and extract and extrude the living waters, turn its irriguous calcium into plates and pipes of pure crystal. Far beyond the summer houses and the marsh grass, beyond the fetch of the casual yachts, beyond the garden and the pink shell, this small seaward arc of calcium's cycle turns without us, indifferent and distant, without our consent.

It is remarkable that the delicate and the small can hold such power, that forams and coccoliths can redress the work of entire rivers. But the world is constructed in this way. Maybe a hundredth or two hundredths of an inch in diameter, the foram is but a tiny marine plankton. Depending on how you look at it under the microscope, it appears bulbous and doughy, and, under high magnification, pitted. But if you look at it straight on, down the vertical axis that runs through the heart of the shell, its geometry at once seems precise and regular. What you see, when the microscopic protozoan is turned toward you just so, is a spiral, more marvelous perhaps than any wentletrap or periwinkle, because of its slightness. The drifting, current-bound foram is an equation afloat on the tides, a gnomon inscribed in calcitic stone.

Even more important in the marine chapter of calcium is the coccolithophore, the single-celled photosynthetic plant that is believed to be the most prolific of the calcite-producing organisms. Armored in plates and ovals, which are themselves nuanced arrays of calcite and

trace organics, coccolithophores bob like buoys in the bottle green of
the sea and fix at their protoplasmic center wind-mixed molecules of
carbon dioxide. Then, in the hidden exchange of green plants, they
transform those molecules in a pierced instant to living matter and to
molecular oxygen. In this way, over the vastness of the sea, dissolved
calcium becomes calcite, and oxygen is set free. Water turns to stone
and the atmosphere is renewed.

I imagine the sea as it was on that night, on the night of the Great
Snow, when in silence and darkness I lay under the eaves, the six-sided
flakes of snow piling higher and higher upon the roof, on all of the
rooftops of the town, over the domes and spires of the hushed cathedrals. You could dream of eternity then: of time without reckoning,
without number, without shadow or moonrise, the sliver of moon
caught in deep cloud, the clock-sounds muffled into background hum.
In the calcareous sediments of the ocean the same eternities gather in
the slow settling of coccolith and foram, sphere upon delicate sphere,
spiral upon Cartesian spiral, humped and layered under the lightless
eaves of the sea. The root-hair prying of calcium from rock, its overland journey through the flumes and valleys of the world, its joining
with carbonate in the patterned blooms of a billion shells, ends in this,
a great "snowfall," as Rachel Carson had called it, in which the ages of
the Earth lie recorded and fixed.

▼ ▼ ▼

In the lab, the burette hung clamped and sparkling above the
Ehrlenmeyer. The calcium solution in the flask—a delicate rose
against the white porcelain stand—trembled a little under the first
drop of titrant. Concentric waves of translucent pink crested outward
and broke against the silvery glass. During the titration for calcium,
the murexide indicator underwent a striking color change from pink
to purple as the calcium ions were withdrawn from it, were enfolded
by the outreaching oxygens of the EDTA molecule. Like other indicators, it brought to the benchtop the hues of New England gardens
and beaches and sunsets over summer seas.

I had finished standardizing, had computed and recorded in my
notebook, when I heard the doors of the Biolab opening and closing.
Shortly Varner appeared and said, "How about some breakfast? It's
almost seven."

The Messenger
and the Disk

The Koolau Range was brown and sawtoothed under a brilliant Honolulu sky, the shield of the ancient volcano worn into ragged river valleys by eons of soft island rains. Wanda, Dana, Kate, and I had gone shopping in the bright little stores of Ala Moana. Among all the vendors with their coconuts and their hibiscus leis, the swirls of colors and aromas and tourist sounds of the morning, two small items had caught our attention. One was a piece of pyrite, the smooth-faced, silver-gold mineral that iron and sulfur make in the dark anoxia of marine sediments. The other was a short, round stick mounted by a thin wooden blade that resembled a propeller.

We put the shiny pyrite into a dark place, into a purse or pocket, to be carried home and admired later. But the wooden blade, by its very shape, called to the blue sky-edges above the mountain, seemed poised for flight. I remember placing the stick between the palms of my hands and moving them back and forth until I could make the blade disappear at will in the sweet blur of pinewood. Outside, my daughters said, "Let it go, Daddy. Let it go," and so I spun it one last time along the length of my fingers and released it to the air. Free, it climbed into the sky, higher than I could have imagined, light wood against the dark brown of the mountains, as my daughters cheered its ascent.

▼ ▼ ▼

I WAS NOT EXACTLY DREAMING THIS, but imagining it in half-sleep as I awakened into the dark room. The night before, I had laid out my clothing in a neat little pile at the foot of the bed. So many things, so carefully placed there: the mukluks and thermal underwear; the red and black checkered shirt; the two pairs of woolen socks; the many-pocketed windpants with their slick outer fabric; the suspenders and leather gloves; the Swiss Army knife with its bristling scissors and spoons and blades and screwdrivers all neatly folded into the red body;

and the red parka with its hood and fine hair. Before I dressed, I showered in the way that you shower at McMurdo: a blast of cold water; then soap; then another blast to wash it all away. The whole thing took no more than a minute, but I felt clean and ready to meet the lakes and the glaciers, ready to set foot on the continent.

Fog had rolled in from the south, had spread over the Sound and the sea, obscuring everything beyond Hut Point. The small fish huts, which were only a mile offshore, were being swept over by the edges of mist, but were still visible like dots of color on some Minnesota lake. I walked over to breakfast, nearly certain they would cancel the flight. If you couldn't see, you didn't fly.

At the mess hall I sipped coffee, wrote in my journal, tapped my feet to the morning rock music that drifted in from the nearby radio station. I ran my thumb under the unfamiliar suspenders, pulled them out and let them spring with a *thwack!* against my chest. Occasionally I glanced up over the rows of neatly lined tables to the colorful murals of the seasons that decorated the wall: mud roads for springtime, iris and daisy. I returned for a second cup of coffee, moved my finger around the rim of the cup, rocked the plastic chair onto its back legs, and watched the sleepy seamen fill their trays.

We had agreed to meet at seven-thirty at Building 73, which is only a short distance up a cinder road from the dorm. When I arrived, Mike and Walt were already slinging crates and boxes and stacking tents outside the barn-wide doors. Varner stood there smoking a cigarette, staring over into the fog from which the valleys were beginning to appear. I entered through the small door and walked upstairs, back through the rows of shelves with their canned peaches and peanut butter and freeze-dried stroganoff. I filled cardboard boxes with packages of oatmeal, instant coffee, and chocolate bars, gathered loaves of freshly baked bread and cheeses that had been carefully wrapped to keep the moisture in. Peas, rice, a few fresh vegetables—"freshies" as people here called them—that had been flown in from Christchurch earlier in the week were crated downstairs and placed alongside the mountain tents and the sleeping bags that lay in bundled rolls in the dust.

I went down to the Biolab and moved our field supplies from the foyer out through the double doors and onto the road: the water samplers in their shellacked pine boxes with little slots inside for the nickel-plated messengers; the conductivity meter with its spool of dark wire; the blue peristaltic pumps and their rolled Tygon tubes stored in heavy battery boxes; the coring device and the bottles of ultrapure nitric acid hidden in inch-thick Styrofoam. I moved them out from the shadows of the Biolab where they had sat all night under Curtsinger's somber photograph of the *Hero*, into the rising mists of a McMurdo morning.

The pile was growing; it was up around my waist now. A Scott polar tent—eighty pounds of green canvas—had been added, looking in collapse like a dour beach umbrella. Mike had brought back cans of "Mogas" for the jiffy drills and "white gas" for the Coleman stoves and for the explosive little Primus. Walt had collected two drill bits and six rugged metal extensions, so that we could cut through the twelve-foot-thick ice covers of the lakes. A box containing plastic dishes and aluminum coffee cups and knives and forks and blue-tipped wooden matches sat beside the Scott tent. Off to the side, just where the cinder hill near Building 73 slides off down wooden slats toward the Berg Field Center, Varner stood with his arms at his sides, looking across the Sound at the changing prospect, at the glaciers emerging slowly from the thinning haze, at the hundreds of miles of glinting quicksilver along the mountains' base. "Looks like we're going to fly," he said, in a low voice that registered disappointment. "I'd give it an hour."

It was about nine when Stephanie came down from the second floor of the Berg Field Center to say, "Everything's go. They want you down at the pad ASAP." We loaded up the red truck, swung it over the hill to the Biolab, piled on the field equipment, and headed off for the helicopters. The ceiling had risen and you could see the peaks of distant mountains beyond the Sound. In the foreground were the Hueys.

I remember a spring evening back in Ohio when the wind was blowing with ominous strength out of the west. The sky had turned pale yellow and there was tornado written all over it, even though the sirens had not yet sounded. Out of the maples in the front yard and down from the empty lot that bordered us, there came a whirring sound like soft wings, the wings of hummingbirds, beating the air. I looked up and saw, sliding and curving toward me, a blizzard of maple seeds. I picked out one and watched it glide from high above the rooftops, downward in a twisting spiral. It had traveled a few hundred feet, maybe more, when it finally came within my range, and I reached up to grab it. But as I did, it lifted again, the single propeller seemingly defying gravity, pulling the seed upward and over my head to continue its journey.

In the maple seed, perfected in the distant ages of deciduous trees, and in the whirligig, handed down from the Chinese through twenty-four centuries of human time, the elegant map of the helicopter lies revealed in miniature. For in the pitch of the blades, in the way they meet the wind and, in meeting it, rise, if for only a gleeful instant, is much of what we need to know about vertical ascent. We need a machine "that will act like a hummingbird—go straight up, go forward, go backward, come straight down and alight like a hummingbird," said Edison. And it was not long, a half-dozen decades, maybe, until we had such a machine honed to such perfection that the pilots came here just to fly it. It was the UH-1A Iroquois helicopter, the Huey.

The crew chief slid open the side doors, pulled them along grooves. We hauled down the sleeping bags and tents from the truck, stored them down under the canvas seats of the Huey. We loaded in gas cans and ice sleds, meters and drill bits, Tecumseh engines and the two long wooden boxes containing the Kemmerer bottles. The cargo bay was filling with dark shapes. When everything had been loaded and the Scott tent had been roped to the skids, the crew chief called us to board. Varner, who had been standing off to the side in some guarded circle beyond the reality of the moment, a lighted cigarette in his hand, moved toward the crowded machine as if he were sleepwalking. His face was nearly expressionless as he looked into the body of the helicopter, where Walt and Mike and I were strapped into our seats, our crash helmets on, our feet lodged against the sleds, our knees tucked up into our chests. When he boarded, and the chief had closed the door, Varner continued to move his hands toward his lips as though the cigarette, which he had extinguished and stuffed in his parka, were still settled in his fingers. Finally we were locked in, the doors closed, the mission to Lake Miers about to begin.

Miers was the first lake of the season. It was a test of the equipment and, more, a test of how we would work together in the field. We wanted to know how Miers differed from Vanda (if it did), how it evolved, how it responded to nitrogen and phosphorus. We wanted to know about the metals. The answers would come in time. Now we needed the samples, the waters beneath the ice.

The first noise was the whine of the engines and then the rhythmic *thunk, thunk, thunk* of the giant blade as it came up to speed on the bolted shaft. The Huey seemed to strain for an instant, made little upward thrusts, hopping motions, a small uncertain bird, before it rose slowly, hovered an instant over the blowing volcanic dust. Then it moved forward, gathered speed, banked as it neared Observation Hill, then turned and headed in an ascending line outward across McMurdo Sound. Below lay the sea ice with its flag-lined roads and its dozing seals like black threads, the tracks of cloud chambers, along the pressure cracks. Directly ahead, in a quilt of whites and soft grays and browns, the mountains and valleys of Victoria Land could be seen as a faint apparition.

There was the sense of not really moving, of just hanging up there, trembling, rolling a bit, being punched by the wind. I was looking into the bulbous molded plastic of headgear, radios knobbed out around the ears, the curved wires of a speaker like a lineman's face mask, and on the back of the helmets the names of the pilots. It was all noise, this huge blade cutting the air not more than five feet away, stuttering a little so that I said, "Don't drop, not now, not yet," and it went back to its old rhythms, air over steel lifting and lifting above the pressure-cracked ice. I was looking at the controls, sticks and levers and pedals and switches and little signs that read ROCKET and MACHINE GUN and thinking about this very ship coming in over dense forest, the doors open, the chatter of machine-gun fire, the Huey twisting as its 2.75-inch rockets headed downrange into the clearing of bamboo and straw, into the scattering of old men and dark-eyed women with their oval faces uplifted, running. I was thinking of Hall and the lush green jungles he had visited before he came here with Benoit—the fronds of giant ferns and palm leaves folding into soft night air on the banks of torpid rivers, and the rice bending over water. "Swords into plowshares," Hall had once said. "I guess that's the way it should be. Swords into plowshares," in that southwest Virginia drawl. And then *slap!* I raised my head into the emptiness of Antarctica, 40 million years of scoured rock and ice, and felt us shake, and there was turquoise light streaming in from the panels above the pilot's head and the co-pilot was saying, "Hey! How are you beakers doing back there?"

The Huey came in high, wheeled above the mountain glaciers and the surface of the lake, and settled onto the blowing sand of the streambed. The blade, now rotating at half speed, cut the air just a few feet above our heads. We unloaded, moving in and out from the machine in a low crouch that was filled with respect. I was breathing heavily, panting with the boxes, walking bent through clouds of my own frozen breath. When the supplies had been moved to a safe distance, we threw ourselves prone on them, lay with our arms wrapped around the equipment crates and food boxes. As the Huey lifted and sidled, a great propwash, like a tornado wind, swept across us, throwing sand against the upturned hoods of our parkas. I closed my eyes, turned my head away, buried it behind a wooden footlocker. The helicopter rose and darted down valley on a low course, its skids barely clearing the moraines.

We stood up and waved, threw our thumbs in the air. In our waving the Earth grew still, the way it is when you're left on a desert road, when there are no sounds. We were performing magic, casting a spell from the cargo. In a few seconds the Huey had disappeared, its exhaust gone into the grayness of the valley.

Mike and Varner set up camp. Walt and I freed one of the ice sleds from beneath a pile of equipment and began to load it for the haul out onto the lake. Under the white canvas surface, we roped in two drill bits and fifteen feet of extensions. We loaded both engines and a can of Mogas and carefully laid on a few pieces of scientific equipment: the conductivity meter with its spool of dark cable, and the Kemmerer bottle with its lines and messengers. Then the two of us grabbed the long rope of the sled and tugged it down into the stream channel and over the grinding, resistant sand toward the lake. It took us nearly ten minutes of heavy pulling, the ropes taut against our shoulders, to make it just to the ice. Once there, though, the sled seemed to find its medium. It pulled easily, almost without a sound. We followed the path of least resistance, hewed close to the shore, moved downlake toward the deepest spot. Walt said, "So much preparation. I feel like we've been gone for years. Remember that place we left? Ohio?"

After half a mile we turned the sled inward across the ice, toward the lake's center. Things began to change. The ice surface became rough and dunelike, its shapes and textures the cut of winter winds. You could see tiny barchans of hard resistant ice, their horns pointing downwind; and longitudinal dunes, with razored edges, meeting at a point a few centimeters above the surface. In the hollows and recesses, sand and pebbles blown in from the valley had accumulated in random patches or, sometimes, in small drifts in the wind shadows of stones.

The monotone of steel runners soon changed into the din and clatter of the loaded sled as it pulled over mounds and sand traps and wedged itself against rigid stalks of ice. Toward the lake's center, we passed boulders and coarse sediment and high cones of ice that, in places, rose five meters above the surrounding terrain. Over the scraping and hissing and metallic clanging of the sled, Walt shouted, almost inaudibly, "What are those rocks doing here?"

In another five minutes we had reached the deep site. Walt grabbed the motor and the ten-inch-diameter bit from beneath the canvas, slid the bit onto the shaft, and fixed it with a cotter pin. We settled the sharp tip of the blade into a small crack, so that the whole assembly now stood at chest height and perpendicular to the ice. I pulled hard on the cord and the engine sputtered a little, then died. I tried it again and again, but it only coughed. Finally it wheezed into silence. The wind was blowing harder now, and the temperature had fallen under the drift of cloud moving in from the Ross Sea. I fastened the hood of my parka tight under my chin and moved my fingers around, searching for whatever warmth might lie in the tips of the woolen gloves. We stood for a moment, let the engine have its rest. The shallow, snow-veined hills curved in from the distance, held us there as in a gray chalice, among the stones of the lake.

I pumped on the small diaphragm to prime the engine. Then I opened the choke wide. This time it started, roared into existence. I eased forward on the throttle and slowly the blade began to turn, began to pull forth the powder of the ice, began to spiral it upward in dry heaps around the hole. The engine became warm and then hot and we were breathing in its exhaust as we drilled. We must have worked for two hours, adding extensions, beating on the cotter pins, driving them through the cold steel, setting up the rig four times as the hole deepened, neared the water that lay below. I threw back the hood of my parka, wiped my eyes, removed my gloves. Cold as it was, there was sweat rolling down my sides deep under the heavy clothes. The torque of the engine was turning us clockwise, a slow two-step around the hole.

The drill wobbled as it spun, wobbled along its full length of five meters as though it would soon tear itself apart. We were bringing up water and ice and slush that climbed the sides of the hole like a hive. Then it happened! A mad spinning of waters, a vortex sweeping over our feet as the bit carved out the last layer of ice, broke through it with its sheer weight, and, still spinning, dropped through into the lake and pulled us down to the ice with it. We sank to our knees with the engine, cut it off, held it safe just above the watery hole. Then we stood up and extricated the rig from the ice. Below us the water bobbed and sparkled and effervesced like a fresh spring that had never seen the light of day. I cupped my hands, reached in, felt its coldness against my palms, and brought it streaming and dripping to my lips.

I was drinking water straight out of the lake, letting it roll down my throat in cold fingers. I was drinking it greedily, gulping it, savoring its taste which was the absence of taste, which was pure sensation, of something going down and wetting and tingling, freezing my throat, then passing. My eyes were shut tight. I was kneeling there, my knees locked into the deep ice, aware of only these thick arrows of cold that were shooting down inside of me. Walt asked, "What's it taste like?" and I answered there's nothing it tastes like. Nothing. It just runs into you and you have no fear of it. It's clear as crystal going down. It's water.

Walt and I had a short lunch, some cheese and the freshly baked bread we had packed at the Berg Field Center. There was wind blowing in from the Ross Sea, cutting up toward the Miers and Adams Glaciers. These were wound around the sides of a pyramid-shaped mountain that seemed to be holding back the thick ice fields that rose in the west. Walt's mouth was dripping fresh water, and it was beginning to freeze in a thin mustache on his lips. "God," he said. "I just drank water. Straight out of a lake. I've never done that before." He wiped his mouth and little crystals glittered on his gloves.

It sounded odd. You live in the world all these years. You're surrounded by water. But you never drink it, don't know what that's like. Water, water, fresh water. But not a drop to drink. Not if it's near the Earth. Maybe you open your mouth wide to the rain. Everyone's done it. A cloudburst and your head goes back behind your shoulders. Maybe you sink to your knees. But now it stings a little, doesn't taste right. It stings and you do it less and less. You learn not to do it at all. Walt said he had walked around Walden Pond one evening a few years ago. Had felt a great thirst come over him. His throat was raw and scratchy. He thought about Thoreau living in the one-room cabin up from the lake, getting his water down there under the stars. The pond was a clear well. You could see the sky in it. You could still see the sky in it. Walt saw it. But he wouldn't drink. He went into Concord instead, found a bar and had himself a cold beer. It was like that almost everywhere. "Maybe that's the great fact of modern history," Walt had said. "You can't drink the water."

▼ ▼ ▼

Our first measurement after lunch was a simple one. We removed the conductivity meter from its carrying case, plugged in the stainless-steel jack of the cable, and lowered the cylindrical probe into the hole. We began at the surface, where the meter read 78 conductivity units. We lowered the probe down to two meters, gave the cable a little jerk so that the electrodes would come more quickly into contact with the lower water mass. We let it sit at that depth until we thought it had come to equilibrium. Then we read the meter again: 85 conductivity units. We did this for an hour, slowly running the cable into the depths of the lake, recording our data in the little clothbound notebook as we went. When we finished, I drew up a rough graph of the conductivity profile and showed it to Walt. The wind bent it in his hands, doubled it over. He had to put it down on the ice. Put his knee on it. "Those are awfully low values," he said. "Looks to me like fresh water all the way to the bottom. A little stratification, but not much. No brine, that's for sure."

It was such a straightforward measurement, so seemingly ordinary that if you didn't force yourself to think about it, you could easily miss how truly incredible it was. Conductivity was the nineteenth-century physical chemist's gift to the oceanographer and limnologist, to anyone who studied the world's surface or subterranean waters. From a single number that tells you how well or how poorly a parcel of water conducts an electric current, it is possible to estimate roughly how much salt is dissolved in it. And from this it is possible to say something about the water's history and even about its future. Oceanographers, knowing the conductivity, temperature, and pressure of water masses, can predict their trajectories, tell you where they will go tomorrow, where they will be next week.

The Swedish chemist Svante Arrhenius first explained why salt-bearing waters could conduct an electric current. To account for what he had observed in the laboratory at Uppsala, Arrhenius had to invent the concept of ionic charge. The problem in the late decades of the nineteenth century was this: It was known that neither pure salt nor pure water was a good conductor of electricity; both resisted the passage of an electric current. And yet if salt were added to water, slipping as it did into invisibility, the resulting solution, which looked to the eye no different from ordinary water, was an excellent conductor.

In 1883, after years of experiments and dozens of penciled notebooks, Arrhenius imagined that when sodium chloride disappeared into the clarity of liquid water it became something entirely new. Not merely small, invisible units of sodium chloride, not merely sodium chloride molecules, but rather charged particles: sodium ions and chloride ions, swarming and bristling with new identities. In aqueous solution, sodium and chloride happily part company, disassociate, and become *ions*, literally "wanderers" in a watery universe of motion and electrical charge. The conductivity meter, with its long cable and its weightless needle on the white face of the scale, was one link between ourselves and the ions that lay below.

Walt and I had been on the lake for many hours, though exactly how many I wasn't sure. The weather had gradually worsened, and snow was beginning to fall in large flakes out of the dark clouds that made their way up valley from the sea. "Grim flotilla," Walt said, looking at the sky. We loaded the sled, covered the instruments and the engine with canvas, tied everything down, and headed back to camp.

That evening we all walked up toward the glaciers. There were mummified seals along the streambanks. Grains of sand were turning in the dead eye sockets. The seals were desiccated and their skin had shrunk clean of the bones, exposing their whiteness. "Jesus," Varner said, "Nothing can live here." He rolled a crab-eater seal over on its back with his boot. "It's been turned to stone," he said. Underneath, it looked perfectly preserved. As though it had died just yesterday.

Mike and Walt were moving on up toward the glaciers ahead of us. The glaciers were a mile away. Maybe they were ten miles away. You just couldn't tell how far off they were. There was nothing to judge distance against. We were walking in the dry streambed. You could see sand and rocks, but they were gradually being covered with snow. In spots there were little clumps of black and dark brown, but you had to look carefully to see them. Varner bent down, removed his glove, pinched one of these between his thumb and forefinger. "What the hell," he said, drawing the thing up toward his face, as though he had pulled something strange up from his garden in Akron. "It's algae."

It was a cyanobacterial mat, a black growth clinging to the stone, above the stone. Epilithic, the biologists said. These were not uncommon in the streams around here. Black little mats hugging the rock, little tufts a few millimeters thick, anchored there tight enough, almost bonded with the stone so the wind couldn't get them. "Life," Varner exclaimed, holding the dark mucilage an inch from his nose. He was smiling. He seemed genuinely cheered by this.

The mats were freeze-dried cells. They were overwintering, dormant, shut down. Fast asleep in the stream channel. But when the water came, the first trickle out of the glaciers, the first wetting, they would begin to awake, to photosynthesize; to take carbon out of the air and turn it into large molecules. They were not insignificant. In the drama of the Earth, they were major players. Three and a half billion years ago, the cyanobacteria had mastered the seemingly impossible trick of breaking the strong double bonds that bind oxygen to carbon in carbon dioxide. They had found a way to turn inorganic carbon into organic molecules, to make intricate compounds by the thousands. Gathered in lipid envelopes, and using only the energy from the sun, these cells could assimilate carbon and a few other common substances and turn them into the peptides, polypeptides, and polysaccharides necessary for life. And they could replicate themselves. But best of all, as waste products they produced molecular oxygen. It was tossed off almost casually, first into water, then seeping and bubbling, leaking into the air. And the molecular oxygen drifted with the winds, slowly changing the mix, changing the winds, changing what they were. God's chemistry: carbon was drawn in and fixed; oxygen was set free. Eventually the blue-green algae would change everything.

We were all kneeling in the streambed. Snow was falling. We had our hoods up around our faces. "It's the Archean Age," Mike said. We're in the Archean. Stone, mountains, valleys. Not much else. And blue-green algae." Then he looked up at Varner and said, "We're back a long, long way."

It took us an hour to reach the foothills, up where we could clearly see the arête rising above the ice. Mike started to walk faster. He walked straight to the snout of the glacier. Mike, the glacial geologist who knew every moraine in Ohio but who had never stood face to face with the agent of their creation.

The glacier was nearly vertical, a white wall shooting up from the valley floor, a hundred feet above his head. Blue ice. Turquoise ice, smooth as jade. We all stopped. You could hear the glacier creaking and moaning. There were caverns and hollows cut in its face. Blocks of ice lay broken at the base, shattered, as though they had fallen from a great height. You could hear the wind moving through fissures and hollows, sighing, making little notes. Inside the ice there was the flex of sheet metal, dry chunks of ice tinkling down narrow cracks. These sounded far away, as though they were coming from a distant room. I looked over and noticed Mike had removed his gloves. His arms were stretched above his head in front of him, like a sleepwalker. He placed his hands against the glacier, his fingers spread. He was reaching higher and higher against the wall, on tiptoe in the moonboots. You could see the naked hills of the Miers Valley, the mountains to the west. And Mike. No one spoke.

▾ ▾ ▾

When I awoke the next morning, the wind was still flapping the tent walls, though more gently than it had all night. The orange fabric sucked itself outward, billowed almost, and then collapsed with a sigh. I must not have rocked down the tent well enough, because the walls seemed closer to me than they had been before I fell asleep, nearly touching my face. My camera swung from the tent's low ceiling above my boots, which I had left at the foot of the sleeping bag. Somehow I had rolled off the foam mat and was feeling jagged rocks protrude up through the floor. The flight bag lay unzipped to one side, where it was spilling socks and turtlenecks and spare issues of thermal underwear all around. There was a tall red notebook, which I had purchased in Christchurch, lying open next to me; a bottle of bourbon tilted precariously from one of the pouches overhead. The parka, which I had used as a pillow, now lay piled over in the corner. I unzipped the sleeping bag, propped myself up on my elbows, and squinted down at the tent's oval portal, which was tightly roped shut against the wind. As I breathed, the moisture of my breath condensed into patches of fog. With my knit sailor's cap still stretched over my ears, I crawled down to the opening, undid the ropes, stuck my head into the air. The air seemed to congeal in my nostrils. For an instant it froze them shut. I awoke instantly.

The valley was wreathed in a gray cloud of snow. The little string of clouds that we had seen making their way along the mountains last night had arrived at our camp and were now spread out in a uniform pallor that hugged the Earth. Where snow had mixed into the cracks and fissures of stone and settled in among the sand grains, the valley floor had the appearance of muslin. In places there were dry crusts of pure white snow that the wind had trapped and packed hard in depressions of sand. Down valley from the cluster of tents, Lake Miers appeared as another shade in the infinite gradations between white and gray. The sides of the U-shaped mountains through which we had flown yesterday had become only shadows. I took hold of the glass thermometer that hung by a string from the tent post. It read twenty degrees below zero Celsius.

I crawled back into the tent, put on my boots, and then crawled out again with the parka in one hand. Outside, I finished dressing, stretched my arms toward the glaciers. Walt was walking up the streambed from the lake. Varner was already dressed. He had set up two footlockers as tables and had started the burners of the Coleman stove. As I walked toward him, he was filling the little Optimus with white gas. After he had pumped it up, he struck a match and placed it over the grill. But the match blew out in a draft of wind and vapor before the stove could light. He tried it again, still with no luck. And then maybe ten times more. Finally, in frustration, he threw a lighted match into the stove. *Whoosh!* It burst into flame, began to consume itself in flame like a burning bush. Yellow spikes of light leaped across the grill and ascended into the air, carrying with them bits of glowing carbon soot. The Optimus roared its heat into the freezing November morning, and then it settled down into a quiet blue pool that was almost invisible. I turned to Varner and asked him how he had slept.

"I'm not sure that I did," he said. "I dozed, but I never had a sound sleep. Every time I'd start to, I'd hallucinate." He looked down the valley toward the lake, scooped up a handful of pebbles, and continued. "Once—it must have been two or three o'clock—I was just dozing off again when I heard this car come driving across the lake ice at a really high rate of speed. It screeched to a stop maybe a hundred feet from the tent. And then the door opened and then it slammed shut and I heard this woman in high-heeled shoes running up the walk to the tent to knock on the tent door." Mike was looking at him with a puzzled expression and then a slight smile. "Wishful thinking," he said.

Varner continued as though he had not heard Mike. "And I just lay there and thought: This is ridiculous. There's no car within twenty-five hundred miles. Women don't wear high heels in Antarctica. There's no path to the tent. The tent is just sitting here on a pile of sand and stone. There's no door on the tent. Who the hell is knocking?" He was looking into the fire. It seemed now that he was talking to himself.

"I was afraid to look outside," he said. "And finally I managed to look out through the hole and, of course, there was nothing out there except the lake. Just the empty lake."

Mike was still smiling, bemused. "I had the best sleep of my life," he said. "I'd still be over there if Walt hadn't shouted in at me."

Walt arrived carrying pots heaped to overflowing with freshly chipped ice. Tilted at an angle, the ice lay in large sheets, like panes of glass or crystalline gypsum, above the aluminum rims. He placed the pots on the fires Varner had made, and soon the ice began to hiss and pop and show little beads of meltwater down where it touched the metal. The process of turning ice into water was very slow. There was heat coming only from the fire, only from the small area of the grill beneath the pots. Everywhere else, heat was wasting off into the valley, off into the vast heat sink of the continent. The Optimus roared and roared and you could see sweat beading on the ice. But it seemed it would never melt. Half an hour passed before we had water. More time after that before the water boiled. We sat there talking about the lake, about the thickness of the ice and how long it had taken us to drill through it, and about the lake waters with their low conductivities. Mike and Varner talked about their work on the camp, how the tent pegs wouldn't hold in the loose sand, how they had had to find boulders to weigh down the edges and anchor the guidelines.

The radio had surprised them, too. After two hours of trying, they hadn't been able to reach McMurdo. Just static and high-pitched squeals coming down like banshees from the upper atmosphere. But Pole Station, eight hundred miles away, had picked them up right away and relayed their message, that everything was okay, back to Ross Island.

As I watched the last crystal of melting ice disappear from the surface of the heating water, I was thinking about a question Henry Frank had once asked. It was over coffee, during one of those conversations he loved to have with his graduate students and postdocs before the morning's work began. Frank was seventy then, but he never missed a day in the lab. And his excitement about his research was as intense then as it had been twenty-five years earlier, when, as a young missionary returning from China, somewhere in the South China Sea, he had had the revelation that liquid water had an unusual three-dimensional structure that explained many of its physical properties. The structure of water had become, from that moment on, the passion and quest of his life.

"Why is it," he asked us, "that ice has such an unexpectedly low heat of fusion?"

It was a curious question because, of all of the hundreds of thousands of solids known, ice has the highest latent heat of fusion. In other words, it takes more heat energy to melt a mole of ice than to melt a mole of anything else. Except ammonia. It takes 1,440 calories to melt a single mole—eighteen grams—of ice. In Miers Valley, we knew this in our shivering bones, as we stood there around the pots and footlockers, desperately waiting for water. I knew this number back in Pittsburgh, too, and so I said to Henry Frank, "But it's not low; it's anomalously high. Because of the hydrogen bonds. Because of the energy it takes to disrupt all the hydrogen bonds in ice."

"You're right, of course," he said. "Compared to other substances it is high. But look at it another way." He was smiling patiently as he backed toward the board, a piece of chalk uplifted in his hands. "How many hydrogen bonds are there in a mole of ice?" By the sparkle in his eyes, I could see that he had already entered into the crystalline lattice, in there among the invisible threads of the hydrogen bonds, like some kind of tiny Maxwell's demon.

Gary, a graduate student sitting on the chair next to me, looked up from his coffee and said, "This shouldn't be too tough. We know that in a crystal of ice, every water molecule is bonded to four surrounding molecules, so there are four hydrogen bonds for each water. That means there are four moles of hydrogen bonds." Henry Frank wrote the number 4 on the board.

"That's true," I said. "I mean, there are four hydrogen bonds holding each molecule in place. But there can't be four moles of hydrogen bonds. There can only be two, because there are only two moles of hydrogen atoms. I think we're overestimating by a factor of two."

Henry Frank wrote on the board, "2 moles H bonds." Then, like Socrates leading his students by the hand to some revelation, he asked, "How much energy does it take to break a mole of hydrogen bonds? What's the bond strength?"

Gary had the number on his fingertips and said, "Four thousand five hundred calories."

Henry Frank wrote this on the board, too, and said, "To break all of the hydrogen bonds it would take two times four thousand five hundred, or nine thousand calories. If ice really melted, if all of the hydrogen bonds broke apart the way you might expect them to, then the heat of fusion should be nine thousand calories. But it's only fourteen hundred. That's why I asked you, 'Why does water have such a low heat of fusion?'"

Compared to what it might have been, had all of the hydrogen bonds actually broken during melting, the heat of fusion was indeed low. The numbers he had written on the board—1,440 and 9,000—represented the actual and the possible, what is and what might have been. Between them lay a story, and I knew that his next question would reveal what that story was. "Let's think about it," he said. "What do these numbers tell you about water, about the liquid?"

Gary was quick to respond. "It must be," he said, "that when ice melts, only some of the hydrogen bonds break. From the numbers you have up there, only fifteen percent."

With the light finally dawning, I said, "This whole argument suggests that in some ways liquid water is still very much like ice. It's as if in the hydrogen bonds, the water recalls what it had been."

"Yes," said Henry Frank excitedly, his hand raised as though he were conducting a symphony. "Water is icelike. The liquid remembers its past. In the liquid, there are these microscopic regions, these flickering clusters, in which the molecules form arrangements that are reminiscent of the structure of ice. Even at the boiling point, one-third of the hydrogen bonds remain intact—which is why it takes so much thermal energy to boil water." His eyes continued to sparkle, but he seemed tired as he finished at the board and laid down his chalk. He told us he was going back to his office. "Maybe we'll understand water one of these days," he said, as he walked from the lab. "But it won't be in my lifetime."

Of course, in some ways we did understand. Once you knew about the hydrogen bond, once you knew about the clusters, you were well on your way. For so much followed from these, these invisible things, these little entities. The disjointed facts in Dorsey's book, the strange properties, at last made sense: water's high boiling point, its thermal inertia, the roundness of its drops, the tautness of their skins, the six-sided shapes of snow, all can be traced back to these precise bonds, these "flickering clusters" beyond our seeing. The lightness of ice, the fact that it floats, goes back to these, as does water's color, its absorption of red, its scattering of blue back to the eye.

The water above the Optimus hit a rolling boil, and you could almost see the violent sundering of those last hydrogen bonds as the molecules tore themselves from the liquid surface and rushed upward, condensing above our heads. Large dry snowflakes drifted through the gray valley and settled like parachutes into the pots, where they quickly disappeared. For a moment we had created on this continent —5 million square miles of ice and cold—an island of fire and heat, a thermodynamic oasis of steam and vapor and burning oil and blue flame, the likes of which had never been seen in this spot. Walt carefully poured the steaming water into our cups. We held the instant coffee close to our faces for warmth, and continued to talk.

Walt was saying that he found the wind last night disturbing. "When it blew," he said, "I thought of imminent danger. I just wanted it to calm down and let me be."

Varner nodded his head as though he knew exactly what Walt was saying. Without thinking, he blew on his coffee to cool it down a bit and said, "You know, it's more the presence of the continent out there, over the ridge"—he pointed a scabrous leather mitten into the air— "that gives me the kind of feeling you're talking about."

Mike waved his arm in a circle and said, "But look, Varner, we're working in this valley where you can see everything. Or at least you could if it were clear. And there's nothing. So what's there to fear?"

Varner looked at him for a moment and said, "I don't like what I see. I don't like any of it."

I was still sipping my coffee, but it had grown lukewarm and tasteless. There were sand grains and bits of mica and some dark, unfamiliar minerals—augite, maybe—gathered in the bottom of the cup. I laid it aside and began to root around in one of the cardboard boxes for something I could eat without preparation—a chocolate bar or a scoop of peanut butter or some beef jerky. Varner seemed to want to talk, but there were too many chores to be done around camp, and I was anxious to get back onto the lake. Miers was not what I expected it to be.

After breakfast, Mike and Walt finished stringing the antenna wire across the fronts of the four tents. They had tied one end to an upright shovel and the other to a six-foot length of drill extension. The antenna was nearly fifteen yards long, and it paralleled the ground at a height of five feet. Walt was kneeling beside it, holding the dark handset of the radio. The bulky leather mitten of his left hand was pressed up beneath the hood of his parka, and his head was bent down into the falling snow as though he were trying to envelop whatever sound might eventually come forth. His voice was getting louder and more deliberate as he enunciated, "Mac Sideband, Mac Sideband, This is Sierra Zero Four One. Do you copy? Over." From the headset there came an immediate burst of garbled sound like rocks rolling over sheet metal. He reached for the clarifier and adjusted it, but the sound remained harsh and unintelligible, as though it were a stream of *r*'s and *s*'s roaring and hissing at him out of the cosmos. Not speech. I knew he would be kneeling there for a long time.

Varner sat on a trunk with a journal opened on his lap and a pencil poised above the rustling pages. He looked out across the brittle snow patches down toward the lake, and then back up the valley to the glaciers. He lowered his eyes and began to write. Mike and I loaded the sleds for a full day of sampling. We packed up the Kemmerer, the barrel filter, two dozen polyethylene bottles, a Secchi disk, the oxygen and pH meters, nitric acid and Eppendorf pipettes, and some cheeses and bread for lunch. Walt, having failed to reach Mac Sideband, patiently turned off the battery-powered radio, attached to it a small array of photovoltaic cells that he opened wide to the sunless sky, and walked over to join us. By now Varner, who had stopped writing, decided not to risk the slippery lake surface with his bad knee, was already walking pensively up toward the Miers and Adams Glaciers. He was clutching the notebook and pencil in his gloved hand.

It was about ten o'clock when we walked onto the surface of the lake. But time had become a useless convention, an artifact of monasteries and corporations, a trick to divide the seamless day. Here, with constant daylight, nothing rang the hour. What time was it really? What eon?

We took the same path Walt and I had followed yesterday. The sleds trailed noiselessly behind us over the smooth ice with its dusting of fresh snow. The snow itself seemed hardly to be falling at all, seemed to be hanging suspended as in a paperweight. As we turned in toward the hole, the sleds began to shudder over the rough ice. Between two cairns of heaped stone we could see our red and green marker flags hoisted on bamboo above the hole we had drilled. We were nearly at the sampling site when Mike finally exclaimed, as if doing a double-take, "What! Why are these stones here?"

The stones were astonishing. They lay in heaps and cones across the surface of the ice, usually at a considerable distance from the shore. There were boulders mixed with pebbles and gravels and wind-blown sands, and in the center of each pile was a core of dark ice that stood protected from the winds. Ice-cored drift. Walt had suggested yesterday that the boulders might have rolled down out of the mountains onto the lake surface. But when he looked around at the slopes he reconsidered and realized that they were far too shallow and the boulders far too irregular for this to have happened. "Bowling balls might have made it down onto the ice," he had said with a grin, "boulders, never." These were like the monoliths of Easter Island, and we accorded them the same respect born of mystery.

The level of the water in the hole was a few inches below the ice surface. The hole had frozen over last night, and we had not brought the drill with us. Mike got down on his hands and knees, reached in, and hit the ice with his fists and with the palm of his hand. But it was more than an inch thick and wouldn't crack. Walt walked over to the piled drift and picked out a sharp boulder that he carried back in both hands. From a standing position, he threw it full force into the hole. Where it struck you could see a frosted indentation that looked granular, like sugar. Outward from the pulverized center, little cracks radiated away toward the edges. Mike reached in again, took the rock, and beat against the surface until it collapsed with that whisper that new ice sometimes makes. Freezing water rushed in around his hands and he quickly pulled them out, dropped the stone, and thrust his hands into his coat pockets.

Walt unraveled the nylon line that he had marked off so meticulously back at McMurdo. At one-meter intervals, beginning with the midpoint of the sampling bottle, he had wrapped small pieces of waterproof yellow tape around the line. At five, ten, fifteen meters, and so on, he had added a red band; and at each ten-meter interval he had placed an additional orange stripe. When the first tricolor of tape crossed the plane of the hole, we would know that the sampler lay ten meters below. I scooped ice from the hole and peered down into the depths of the lake. I was looking through a crystalline pipe whose sides appeared undulant and glassy and tinged with blue. Waves of blue glass. Below it, all around, falling away into an azure depth, lay the clearest water I had ever seen.

Walt opened the gray barrel of the Kemmerer bottle at both ends so that water could pass through as it sank. The bottle had a large volume, more than six liters, and was constructed entirely of plastic. With his feet spread so that he stood directly over the hole, he lowered the sampler until its bottom end touched the water's surface. He let it sink slowly, let the water rise up gradually through the cylinder until it finally covered the top. He fed more line into the lake, watched the yellow stripes of tape slip beneath the surface. When the sampler had fallen five meters, he stopped it, squeezed the line between his thumb and forefinger, let it hang there for a while, still perfectly visible, moving back and forth like a massive chandelier. I took the "messenger" from the box, clipped this simple metal bar onto the line, and with a flick of the wrist, shot it down into the lake. When it hit the Kemmerer, the bottle closed instantly and huge bubbles, like translucent fish, welled to the surface. Hand over hand, Walt brought the sampler in, held it upright on the ice, and pried open the top. It was filled to overflowing.

We sampled for twelve hours. The open Kemmerer slipped into the lake, submerged itself with a *glub!* as the frigid waters poured over its lip, down the inner sides, and back into the lake. It fell along the crystalline ice, touched as it sank, gently bounced away, centered itself in the hole. We lowered the sampler, checked the colored markings against the ice, double-checked, sent the messenger, waited and waited and felt for the tug, the tug of the filled sampler stretching the line. We pulled it up as water from the line beaded into tiny droplets, filled the air, and froze as we worked. Into the polyethylene bottles arrayed along the sled we emptied filtered water, six liters from each depth: two liters for nutrients; two for metals; two for major ions. We labeled and relabeled the bottles, wrote in waterproof pen, assigned each a number, and, in one of the little green notebooks, wrote what that number meant, where the sample had come from and on what day. By evening our gloves had become blocks of ice.

Toward McMurdo Sound, beyond the drift mounds, the sky had turned light. Opalescent clouds, suffused with wisps and streamers of fine-spun pink, stretched above the sea ice across the broad opening of the valley. In the middle of this, just where the curve of the mountains reached its lowest point, stood the hills of an island, visible for the first time. The sky to the east was warm and light-filled as though the sun had just set. For a moment it made me think of evenings on wooden decks, of olives and palm leaves, the flatness of a summer sea.

We loaded the sleds, stowed the tightly capped samples under the heavy canvas, and tied them in. There were bits of plastic and Styrofoam chips and pieces of string and Tygon strewn around the hole or refrozen into small lenses in the ice. We pried them loose and picked them up as best we could, until the surface looked much as we had found it.

Before we returned to camp, I wanted to take one more measurement. Out of the prow of the sled, I pulled a disk the size of a pie pan. The Secchi disk was forty centimeters in diameter and divided into alternating black and white quadrants. Beneath the disk a small square was attached as a sinker. On top there was a ring to which a line could be clipped. The idea was to lower the disk into the water until it could no longer be seen. In the limnologist's trade, among "the gear and tackle and trim," this was the simplest of all things—a circle in search of the dusk, in search of that thin stratum of lake where light and dark merged to become one.

Though over the years it has become one of the essential tools of limnology, the Secchi disk was named in honor of an astronomer. In the nineteenth century, the wells of Italy became contaminated. The contamination, it seemed, was related to the turbidity of the water source. The Pope, fearing an epidemic, called upon the famous astronomer and popularizer of science, the priest Angelo Secchi, to devise a reliable means for determining the comparative clarity of well water. Secchi had already pioneered the use of photography as a tool in astronomy, and he had been among the first to obtain spectra of the planets, the sun, and the brighter stars. A master of telescopes and spectroscopes, he was no stranger to the ways of light. Father Secchi accepted this hydrologic challenge. He fashioned a small disk that could be suspended from the end of a rope and that could, within seconds, provide information about the quality and safety of a water supply.

Secchi's disk is used today to determine water clarity, and limnologists have obtained "Secchi depths" on lakes throughout the world. Not surprisingly, there is a reasonably good correlation between the

Secchi depth of a lake and its biological productivity; in general, the correlation says, the more biologically productive a lake is, the more organisms it supports, the more light it scatters, the smaller will be its Secchi depth.

As the weather cleared and the sky became a powdery blue over the eastern valley, I began to lower the disk through the hole. It settled into the water, jerking a little from side to side as it sank. I could see it clearly as the five-meter stripes passed over the ice, and I could still see its black and white quadrants as the first tricolor touched the water. At fifteen meters it appeared small and far away, like a silver coin dangling on the end of a string. In another meter I had lost it. "Sixteen meters," I called to Mike, and he wrote it down in his notebook. The lake was clear to more than fifty feet.

I remembered an autumn day back in Ohio, the trees newly stripped of their leaves, the sky a brighter blue, when Walt and I tried the same experiment in Acton Lake. No sooner had the disk started its descent than it began to blend into the murkiness, began to be extinguished by the suspended particles of the lake. First it became a shadow beneath the boat, then it disappeared entirely. I recalled the Secchi depth as being not much more than a meter.

As I worked, I thought of a passage that Thoreau had written in the winter of 1845. He was recalling what had happened to him at Walden many years before, on a day when he had been cutting holes in the pond ice so he could fish for pickerel. He had momentarily stepped off the frozen surface and had casually tossed aside his axe. But the axe had slid, had found its way into one of the fish holes. Thoreau had gotten on his hands and knees and had peered down into the lake. What he saw was the axe, standing upright and clearly visible under twenty-five feet of water, its helve slowly swaying with the internal waves of the pond. In his journal Thoreau had recorded the clarity of water on that winter afternoon, the quality of light.

We moved the sleds out, turned them west toward camp. Blue sky shot in streaks beyond the glaciers. The lake ice, in spots, had become like glass, and its sculptures bent the rays of evening light. As we pulled, I thought of the messenger: the way it fell along the line; the way it shook the silver bubbles from the lake, shook them into bursting like grapes. And how it closed the sampler around a precise parcel of water—the one that we would come to know in our limited way as the lake itself.

I thought of all the messengers we sent into the folds and fires and distances of the world, into seas and galaxies, into its microscopic hollows and sockets, its very being. Those messengers of our craft and invention returned to us from those places—bights and hinterlands below our seeing, beyond our imagining, pasts of warmth and abundance—returned as emissaries bearing gifts and tidings. Alpha particles transiting the disks of atoms; ionic currents streaming in laboratory glass; starlight and planet-light spreading like fingers through a prism in a dark room; the ladened ships of exploration returning, struggling to return; all the imprints and splinters and shards of bone pressed into ancient rock; heat, pulsed into a cup of ice; stones heaped quietly upon a lake. All of them—light waves, water waves, particles and shells, wisps of energy and reflected sound—all were messengers. The disk itself was a messenger, and the Kemmerer, and the meters all seeking the measure and structure and order of the world.

SEVEN

Stone

I remember the first time I saw a facsimile of the engraving. It appeared in Hutton's Theory of the Earth, *that great eighteenth-century treatise that I associate with the pubs of Edinburgh and with the genius of the Scottish Enlightenment, with Adam Smith and Joseph Black and James Watt and all of the musings and prophecies that came from those cold winter nights under the curling dense coal smoke of the chimneys, all of those portentous reflections on labor and property, the composition of the atmosphere and the flywheels and pistons and efficiencies of engines. Things that men in pubs thought about over stout and Guinness on cold Scottish nights. And somewhere across a thick wooden table sat Hutton with his vision of the great machine of Earth, the great heat engine turning up its mountains, leveling them, washing them to the sea, raising them again, always raising high the roofbeam of Andes and Himalayas straight into the mist and moonlight of far-flung nights, then calling them back, always, in Hutton's time without reckoning, calling them back. And in the engraving you could see these things: On an ordinary road a rider on horseback passes a horse-drawn carriage. There are a few trees bending this way and that, a farm field just beyond the carriage road, and some low-lying hills standing against the sky. So far, just an ordinary scene of rural Scotland. But then, beneath it all, beneath the carriage and rider, the trees and shrubs and fields, lie layer upon layer of Devonian sandstone, the strata growing thinner downward until they give way to a layer of angular detritus. And then, below this, even more rock, beds of sandstone and Silurian shales, some nearly vertical, others bending a little, like rice stalks or spartina, as they rise upward to meet the detrital fill. In the library, I remember holding the book at arm's length, taking the engraving in all at once. Atop so much stone, so much time, atop the dropped sand of braided rivers and the calcites and clays of ancient seas, atop the Silurian reef assemblages with their tabulate corals, and the filigreed fossils of lobe-finned Devonian fish, the carriage and horse, the riders and trees like wisps of hair and fine straw, seemed a mere sprinkling of dust and dust mites, spindly things and wind-blown, strewn across the layered, time-burdened stone. I remember walking out of the library onto the streets of Pittsburgh, with its cobbles and rails and its smoke-dark sandstones, and seeing all of this, and myself, as unbearably light and fleeting, a thin molecular layer clinging to the surface of something that went down and down and down into deep-dwelling places where the wheels of carriages, the clop of horse hooves, the roll of trolleys had never been.*

▼ ▼ ▼

A T MIERS, THINGS HAD BEGUN to come together. After a week, the pumps and meters had begun to work. The collection method for the metals seemed free of the usual problems of contamination. Varner's natural curiosity appeared to be reviving. As long as he kept himself focused on the immediate vicinity, on problems of measurement, on observations of wind and rock and patches of life, he was fine. Mike and Walt were beginning to understand him, to catch his flashes of humor, his self-doubt, his peculiar, patient engineer's way with objects. One evening they brought him a motor from the lake. It had fallen apart in their hands. Springs and little pieces of metal had been shaken from it. It was useless. Varner worked on it for six hours in the tent. He looked like a watchmaker, sitting there surrounded by parts that were skimmed with grease. When he gave it back to them, they took it just offshore. They pulled the cord. It started right up, cut straight into the ice.

He was also becoming more observant. "Look at the sun," he said. "I tracked it yesterday for twenty-four hours. Watched it circle above the horizon. It never sets!" He had measurements of sun angles and temperatures. Of course he knew the summer sun never set here. We all did. But to actually see it, to stay awake for twenty-four hours just to watch it! As he talked, you could almost feel the whole planet turning you, feel it move like a great silent wheel carrying the mountains and valleys, the whole frozen lonely sea as it went. "Three hundred sixty degrees. The whole damn circuit!" he exclaimed. "What would Copernicus and Galileo have given for this, to have seen it all without night? I can't wait to tell my physics class."

Walt had finished reading Medawar's *Advice to a Young Scientist*. He had been reading it since Los Angeles. He had come running out of the tent with the book in his hand. I was on the shore of Miers, looking for the first signs of melt, the first water. But it was still too cold. Still bitter spring. "Look at this," he said. "'This is just the way it is.'" He flipped pages into the wind, found the place, held the book down near his waist. He read the passage where Medawar talks about the way science works, how it is experienced by the scientist not as boring, tedious puzzle-solving, but as a continuing revolution of thought. Little revolutions, ferment, bursts of passion that could carry you right along. How, during any study, everything appears to be in flux. How you never know what you're going to find next. "So much for 'normal scientific life,'" Walt said. "So much for God-fearing, bourgeois contentment. So much for neat little methods written down in stone."

He closed the book. "I know now why I love this stuff," he said. "It's not like the textbooks say at all. I can't wait till the nutrient data come rolling in. We don't understand anything yet."

There was a growing sense of cohesion now. We were entering the work. It was drawing us together. But I still worried.

▾ ▾ ▾

The chopper was pulling up slowly, wobbling a little as it rose, then righting itself, pulling up again along a vertical thread that ran out of the Earth and into the sky. Mike and I were on our way to Vanda to set up camp and begin our search for the traps. Varner and Walt would remain at Miers a few more hours. They were standing off to the side, the hoods of their parkas raised, turned away from the blowing sand. You could see the sand like a fine brown rain glancing off their coats. The window of the Huey was filling up with ice and mountaintops and intense blue sky, and Varner and Walt were turning into tiny specs of color down on the valley floor. The helo's stick was being pushed forward and we were gathering speed over the brown burlap folds of the valleys up toward the glaciers, banking hard above their snouts, turning back east toward the sea, rising to where the outline of Lake Miers could be seen.

From Miers Valley, there are two ways to Lake Vanda. You can fly off toward Cape Chocolate and hug the foothills of the Royal Society Range where it drops off into the Ross Sea. This way you curve along the Strand Moraines and the Bowers Piedmont Glacier past Butter Point and New Harbor and Marble Point—all those mythic place names from Scott's journal. At Marble Point you refuel. Then, after lots of exuberant waves and thumbs-up from the Navy people there, you head due north into the Bay of Sails and inland over the Wilson Piedmont and the Lower Wright Glacier, into the long, narrow fjord that is Wright Valley. Then down the dry-braided channel of the Onyx River to Lake Vanda.

That's one way. On this trip, we went the other. We took the high road, the "sky passage," up among the peaks and snowfields and precipices of the Transantarctic Mountains. The alpine road through heaven. We doubled back and swung over the Blue Glacier, climbed to four thousand feet, up where the landscape is ice-scoured rock and wind-sculpted snow, as smooth as custard. Outside the window I was looking at distant Mount Lister gliding by as if in slow motion, the rock of this twelve-thousand-foot peak wreathed in glaciers and breathing streamers of snow straight into the cloudless sky.

We passed the Pimple and the Camel's Hump and bisected the long finger of the Ferrar Glacier, which extends thirty miles to the west where it joins the Polar Plateau, and thirty miles east, where, near Mount Barnes, it floods into the sea at New Harbor. The Ferrar, a Mississippi of inching, creaking ice, seems to cut through the Kukri Hills as though they were mere butter, the ice stream moving more slowly where it touches rock, more quickly in the unimpeded center. You can see this in the flow lines of the ice, the way they seem to hang back near the mountains and rush forward toward the ice stream's center.

Not long ago, Scott had set out from the *Discovery*, had crossed McMurdo Sound to New Harbor, and had clambered up the tortured ice falls that lead from the sea ice onto the vast Ferrar. It was September 16, 1903, and Scott and his party were down there sledging into the continent's heart. Robert Falcon Scott, the continent's most ardent lover, eight years away from his untroubled death in the tent. Only the wind outside. I half expected the breath of his men to be hanging there, visible as we passed over.

Three thousand feet above the surface of the Ferrar, boulders the size of houses cling precariously to the slopes. The Ferrar was once deeper and mightier than it is today. It had rafted these rocks, like the erratics of the Jura Alps, far into the highest mountains. Scott and his party had seen these boulders, these very ones, the first humans to lay eyes on them, and had understood what they meant—had understood the power of the ice. When ice sheets meet stone, there is no contest.

It may seem surprising that mere frozen water could move stone, but ice is really rock, and glacial ice is rock that moves. Like igneous rock, ice forms from a fluid—"molten" ice is water. And like sedimentary rock, glacial ice accumulates in thick layers. This year's soft snowfall is covered and compressed, the air partially squeezed from it until it becomes denser, becomes "firm," and then, under further compression, becomes granular ice and eventually glacier ice. In a sense, a glacier is metamorphic rock, rock transformed under pressure, recrystallized. Enough years, enough snowfalls, and the "sedimentary" layers of snow build and weigh upon one another until the ice is thousands of feet thick, burdening and depressing the land beneath it, flowing outward from its center, polishing the stone. The thicker the ice sheet, the greater the slope down which it flows, the faster and more powerful will be its action.

Flying through the Transantarctics, you see the work of ice, you see its signatures as a sculptor of the land. As you might attribute a wind-drifting balance of strings and plates to Calder, or a gaunt, striding figure of a man to Giacometti, so there are certain features of the land that only ice could have made. These cirques, these natural amphitheaters cut high in the mountains all around us, these U-shaped hanging valleys and the snow-dusted ridges between them, these are unmistakably the work of ice. But here in these mountains it's as if the sculptor were standing beside his creation, still crafting and chiseling even as we watch. And if you find it difficult to believe, as I do, that Yosemite Valley and Bridal Veil Falls and Plymouth Rock and the crosier-like arm of Cape Cod are all among the handiworks of ice, then the Transantarctics would convince you that this water, turned stone, can do anything. Mike was looking at it all. His face pressed to the window. He turned to me and, over the engines, shouted, "So this is what it's all about. Now I understand."

We crossed over the Kukri Hills and the Taylor Glacier, which is a tributary of the Ferrar, and moved in among the glistening peaks of the Asgards. We were flying at no more than a hundred feet above the terrain, which in my mind, and for reasons that have nothing at all to do with physics or aerodynamics or reality of any kind, seemed a comfortable altitude for a helicopter.

Then everything changed. Near Obelisk Mountain we were shaken by crosswinds. It felt as though the chopper had been punched, as though an invisible hand had come out of nowhere and delivered an uppercut to the nose of the machine just as it began to clear into the valley. We were not moving in any direction, just sitting there shuddering, straining against the wind that was rushing off the plateau. And we were no longer at a "safe" distance above the Earth. We were over the Wright Valley, and its floor was six thousand feet below.

We were falling, descending the steep walls of rock and scree, try-ing to turn into the wind, turn west toward the Upper Wright Glacier. Then the chopper caught and steadied itself, pushed hard into the headwinds. The *thunk* of the blades became louder and more urgent. We were nearly there. The lake, which, from the high ledges of the Asgards, had been a mere droplet, a glistening gemstone of frozen water, was now, a few seconds later, filling my whole vision with its blue ice. We were nearly turned on our side.

In an instant everything changed. We were braking, coming in parallel to the ground, near the western end of Lake Vanda, inching up a little over a small moraine. The crew chief opened the door and there was sand whipping around everywhere and the noise of the wind and the engines filled the cabin. My breath froze instantly on my glasses as I climbed out, the machine hopping on its skids.

There was not much to unload, so they kept the engines running. We ducked under the blades with our gear and tents. I didn't want my knit cap sliced in half the way "Big's" had been, right here, right on this spot. The pilots had placed that cap—both halves of it—on dis-play at the helicopter pad. You noticed it every time you flew. "Big's" hat had been split right across the middle. We gathered everything into a pile and flung ourselves over the mound of gear. There was noise and sand, the deep sounds of flapping canvas. The Huey rose and was off down valley. Then it was quiet—except for the wind. I was standing on the toes of my boots, trying to fall forward into the west, but the wind was holding me upright.

Mike and I tried to raise a Scott tent on a moraine overlooking the lake. We were near the far western shore, down where the lake was deepest. Mike got on his knees and crawled in under the canvas and stood up. He was gripping the pitch pole, swaying with it, trying to steady it. I was pulling out fabric on the outside, at the bottom, rolling the heaviest boulders I could find onto it. Mike was yelling something to me, but his words were carried off like whispers, sucked off into the air. I could hear him staggering around as though he were going to fall at any moment. I tried to get purchase on the sand and rock, working the canvas outward as I moved along the tent's circumference, rocking it down. It took us nearly an hour to set up.

The landscape outside the tent's portal was frozen, motionless. In the distant streambed of the Onyx River, which lay far to the east, there was unlikely to be a single drop of water. The lake itself was all stillness. The mountains rose and fell, angled themselves deep into the morning sky.

After we had rested, we set up the radio and I called McMurdo to tell them we had established our camp site and that we were safe. I checked to be certain that Tim and Dr. Yu would be coming out to the valley on the next available flight. Everything was "go."

I asked whether Walt and Varner had arrived safely from Miers. They had. But the radio operator at McMurdo had a message. He called it "urgent." Over a low buzz in the receiver, he read it slowly, carefully, repeating each sentence, punctuating it with "How copy? Do you copy?" The distant voice on Ross Island was saying over the wind, "Mr. Varner, repeat, Mr. Varner will depart Williams Field on next available flight. Repeat. Mr. Varner will depart Antarctica on next available flight. How copy? Do you copy? Over."

There was no mistake. I had heard it right. "Copy all," I said. "Copy all. Sierra Zero Four One over and out." I had copied everything, but at first I couldn't believe what I was hearing. "Mr. Varner will depart Antarctica." "But why?" I said to myself. "It was all working out."

I walked up toward the Asgard Range. My face was flushed. I unzipped my parka and unbuttoned my flannel shirt to cool down. "Depart!" I shouted to Varner, who was somewhere across the Sound, who was back at McMurdo. "Without doing anything? Without building the damn flume? Everything hangs on our getting the stream flows right." I was talking right into the winds that rolled off the plateau, the gravity-driven katabatics that could blow you over. My words were breaking into pulses of atoms, into speechless sound, into pure babble. "I can't design a flume!" I yelled, as though Varner were standing at my side there among the rocks. "I'm not an engineer. Maybe I can throw some matchsticks into the current and punch a stopwatch. What will that tell us? Who would believe the numbers? You came here to make these measurements. I thought we had a deal." I looked down to the tent, even pointed down to it, down to the radio. "I can't argue with you over that. I can't even order a loaf of bread with the damn thing."

I was near the base of the mountain, more than a mile from the tent, when I finally stopped. The wind was howling and I was suddenly feeling cold. I sat on the lower slopes, below the peaks of the Asgards, and stared down at the lake. It was a crust of ice, nothing more. "Something must have happened," I said. "It must have been the flight back. He must have started thinking about the whole place again."

Across the valley I could see the Olympus Range. It just shot up into the blue sky—six thousand feet of sandstones and shales, 400 million years of time. It looked like a picture I had seen.

The Dais lies almost in the middle of the Wright Valley. It rises steeply from the valley floor then flattens out like a table top. You can imagine it being the seat of some ponderous stone god, or maybe the god of emptiness or the god of eternal solitude. It has that look about it. Like so much on this continent, it was carved by the ice.

Millions of years ago, in what geologists call "Vanda Time," the ancient Wright Glacier began its journey from the frozen interior of the polar ice sheet. It moved eastward toward the sea. An observer, viewing the distant Earth at that time, would have seen only a minor eruption of the ice cap, would have seen it sprout yet another little tendril of whiteness from its center. But a more immediate witness, some sentinel at the top of the mountains, would have seen a slow-motion catastrophe enacted beneath him in thunder and majesty.

The glacier, at its full extent, was more than three thousand feet thick—nearly three times the height of the Empire State Building, as deep as many of the world oceans. At its base, the pressures were enormous. So when it moved, it moved with unchallenged authority, as though the hand of God were holding it like a giant awl to the skin of the Earth. In Vanda Time the Wright Glacier gouged and excavated fifteen hundred feet of bedrock from the area around Lake Vanda and pushed it down valley, off into McMurdo Sound.

What remains of that old bedrock is the Dais. It divides the Upper Wright Valley into north and south forks, and it stands like a weathered medieval fortress above the shores of Vanda. I remember the first time I climbed it on one of those endless treks with Benoit. That was more than twenty years ago. We had been up valley for six hours, just looking around in a kind of indiscriminate way, looking for salt deposits or fossils or desiccated seal carcasses, when Benoit suggested that we climb to the top of the Dais. "Beautiful view from up there," he said, in his laconic Vermont way. "It'll only take a few minutes." Then he went striding up among the boulders, unconcerned with whether Hatcher and I were following him.

At first I wanted to curl up in the hollows of one of the great sandstones that lay at the foot of the Dais, and just fall asleep until they returned. But after a few minutes of keeping pace with Benoit, I realized I had entered into an unusually strange corner of this already strange land. I wanted to continue. Rising up the entire thousand-foot wall of the Dais, and sloping off into a wreath of cloud, was a Bangkok of statuary whose numbers were beyond reckoning. All around me, on either side, above and below, massive sandstone figures huddled together in improbable associations—gigantic, wind-carved monoliths leaning against one another like the remnants of a failed civilization. Elephants with their great haunches lurched toward me out of the rock face. Off to one side were a huge pair of sunglasses and the massive stone head of a therapsid. A hundred abstract pieces, boulders rounded and hollowed into modernist shapes, staggered up the face of the Dais. I worked my way in among these forms, crawled beneath them, stopped to run my hand over their surfaces, caught my breath in the darkened catacombs that looked out from them toward the lake. When had Henry Moore been here, I wondered. Mixing space and mass. Letting space flow right into the very heart of matter.

It took us two hours to reach the top—just as Benoit had promised. We were standing there with thin cloud wreathing our feet.

▼ ▼ ▼

Evening. The wind had died down, but the temperature stood at minus five. The chopper had not come with Tim and Dr. Yu. The pilots by this time were having a few drinks back at the officers' club. There would be no flights until morning. A low sun was breaking through between the peaks of the Olympus Range. There were thin clouds ringing the Dais. From outside the tent where I was standing, the rock seemed almost plum-colored in a way that I had not seen before.

EIGHT

The Lake

Every day began like this: The zero-degree water cupped in my hands or taken straight from the lake, my head underwater, just above the clear stones. Then more water for boiling, for coffee. I would sit on the cold ground by the tent, thinking just how slowly things seemed to move here when you looked at them close up, minute by minute: the water barely convecting, turning its lazy loops in the pot above the flame. But then, how, when you stood back, and saw time spread out, it seemed to be careening away, seemed to be racing off into the mountains to join with all the hours up there that were past.

We live our lives between plains of darkness, Nabokov said. The light is as thin as a laser beam. We pass through it so quickly. And yet in its midst there is the illusion of endlessness. "This cannot end," we say. And the beam, such a fine cut and sliver, opens so perfectly into eternity as we walk that we are lifted by it; for an instant it feels as though we are immortal. I felt that here. Somehow the light would be an expanse so great that the darkness before me was invisible.

▼ ▼ ▼

THE DAYS MOVED SLOWLY NOW. The valley and the mountains had their own rhythms to which we could only bend. Moaning winds up in the Labyrinth, cracking ice somewhere across the lake, distant land-slides, and the coming and going of snow in the high Asgards (the "white jade mountains," as Dr. Yu called them)—these were the things that happened. Not much else. After Tim and Dr. Yu's flight, no other helicopters came into the area. Occasionally a skua gull would catch an afternoon wind and drift in from the Sound, but other than that, noth-ing moved. "Coms," communications, between Ross Island and Wright Valley had been poor all week. The best I could do was to get word out each morning that we were okay. There was no further information on how Walt's analytical work was going, and there was nothing on Varner. I assumed he had left for Ohio.

I went down to the lake first thing each morning. The nearshore ice was still frozen, but it was thinning a bit at the edges. It was easier to punch through now. I knelt on the floor of the valley and hit the ice with my fist. The ice cracked open with a hiss, the way thin ice often does, and then splintered into fine crystals that swam in the blue water. I cupped my hands and drank from the lake as though it were a fresh desert spring. Then I plunged my head into the water and let it rise on the back of my neck. I shivered and pulled up and it streamed through my hair and over my face and down into the cold sand. I was awake. I took a knife from the pocket of my windpants, opened it, and looked at my face in the blade. Then I stared out at the lake.

Maybe it was the wind on Vanda. The *sssshhhh* and click of the sand. Then *sssshhhh* again. Maybe it was just the play of light on the dazzling mirrors of the ice. But for a moment the skates ran on a limpid skin of water. That's why he could glide, tall and elegant, the snap of his cassock in the wind. Why it was so effortless. Why there was so little sound. How he could glide!

It was Indiana, the little valleys of the south covered in birch and maple and the pond hard and smooth. Sunday you could hear the monks singing clear as chimes, the Gregorian chant slipping down as if from the skies, slipping around us, and the little sound of the skates —*sssshhhh* and then click then *sssshhhh* again. Heaven and Earth! Eugene was only fourteen, too young to be away at school, I thought. God and the angels would persevere another few years without my brother. But winter felt saintly. And he was there, a seminarian, servant of the Lord.

Later, when I was also studying in Indiana, we would meet in Indianapolis and board the train to Pittsburgh. Eugene was reading Melville and Henry Adams, plotting how he would get from the small seminary to Harvard, where he would one day study literature. "You must read Adams on Chartres, Bill." And we would plan how we would meet one day in Joyce's Dublin or in the gritty Belfast of Kelvin. This was my younger brother, the kid I had played war with, the guy who'd lobbed baseballs through the neighbors' windows. He was speaking like a scholar. "Paris, Oxford, Rome." He had already mastered Latin and French. Italian was next and it would be a snap. Years later the letters and postcards came: Eugene in Provence, in Venice, in the Lake District. Meditations on Wordsworth and Chaucer, on Silone's *Bread and Wine*, on Newman's *Idea of the University*. Lake Geneva at dawn. Athens at sunset. A young man in love with the world, with the preposterous abundance of it.

From Chartres one year, Eugene wrote that he had been accepted at Harvard. He remembered the very spot in the great cathedral—even the way the light fell, the cold stone—where he had made his decision to go. It had not come easily. He was in his early thirties then, already a priest, already famous in Pittsburgh for his oratory, for its poetry and wit, its weaving of the Gospels with the words of Eliot and Mann and Virginia Woolf and with all that he had seen. He was chaplain to the universities, a frequent lecturer, a celebrity, almost, in the small academic world that was Oakland.

At first, Boston and Cambridge were wonderful. He lived on Beacon Hill, not far from the golden dome of the State House—not far, in fact, from where Henry Adams had spent his last years. He was quickly drawn into the life of the city. Catholic Boston! The American Athens! What else did he need?

We hardly communicated in those years. Polite Christmas cards and birthday greetings, that was all. I was trying to set my own feet on the ground out in Ohio. Wanda and I had just had our first child, I was learning to teach and, almost in my off-hours, was learning geochemistry. At times our visits to the Cape would overlap, but as I recall we were stiff and formal with one another, and too often we hid behind our books. Our sister, Elizabeth, would say, "You and Eugene need to talk. And soon!" It was almost as though we were different people, not the same two. But I thought there was plenty of time. I thought it would all come full circle one day, the way it usually does.

We had come to Vanda to collect water and to retrieve about a dozen traps—clear plastic cylinders, sixty centimeters long, ten centimeters in diameter, open at the top and capped at the bottom. A year ago we had suspended them in pairs from nylon ropes at different depths in the lake, where they hung as waiting receptacles for any particle that should settle into them from above. Back in Ohio I had spent nearly a year anticipating what we would find in those traps once we had pulled them to the surface. I felt certain that the trace-metal story would be spelled out in the chemistry of the trapped particles. As we laid them in place a year ago, almost as we watched, the lake ice had closed in around the lines, locking them and their suspended assemblage into place. The trick now was somehow to bring them back, like tall, thin Houdinis, through twelve feet of solid ice.

During the days at Vanda we sank many holes. Using the ten-inch bits and the Tecumseh engine, we cut them as close to the trap lines as we dared. The ice was still hard and dry and there were no pockets of meltwater to spin up out of the hole. This made the drilling easier than it would have been later in the season, and we were cutting through the ice at a rate of three holes per day.

To recover the traps, Tim had put together a long, L-shaped metal pole that could be lowered through the ice hole. The pole descended in its folded position, but once at the bottom of the ice, it fell open at the hinge. The idea was to twist the pole until the horizontal leg snagged the line from which the trap was suspended. Then the whole device, with the trap attached, was to be hauled up through the freshly drilled hole. It looked good on paper.

Toward the western basin of the lake, where the water reaches a depth of seventy meters, we lowered the "trap catcher" through the first hole. The hinged leg fell open as soon as it had cleared the bottom of the ice. We held the pole vertically and began to turn it slowly, sweeping out a circle in the water twelve feet below. We had turned through 270 degrees when we felt something catch. Tim and I were holding the pole, and we both felt the resistance at the same time. Tim yelled, "Damn, I think we've got it. We've got the line." We twisted the rope around the pole until it felt as though we were trawling with a massive weight. Then, just as we were about to start bringing it up, something gave. We were spinning with nothing on the line. "It got away," Tim said. "We lost the fish."

We lost the fish many times. Finally we lost the pole.

We could not repair the hinge with the few tools we had brought with us, and after a dozen lowerings, the vertical steel bar of the L had been bent nearly in half by the mad winds. We could do nothing to straighten it. The metal trap catcher was useless. I felt as though we had just shot an entire year.

From the western end of Vanda it is about eight miles up to Bull Pass. I started early so that I would have time to clear my mind, review where we were with the project, and be back by dinner. As I left, Dr. Yu was sitting outside his tent, playing around with a single bamboo pole. "Bamboo and rope," Dr. Yu said. "Can make anything."

In spring, the Onyx River winds and cuts along the base of the Olympus Range just where the mountain begins its steepest ascent toward Bull Pass. I sat by the river, which was locked in winter, still silent and without water, a powdery snow over the braided channel, and rested before I began my climb. There were clumps of freeze-dried algae scattered across the shallow pan of Bull Pond, which lies just above Vanda. Cathedral light streamed through the clouds and illuminated the dry pond so that it looked silver and black.

I had rested often at this spot. In the early eighties, when Canfield and I first began to study the river, we had set up one of our sampling stations here. When we started our work on the Onyx, we had no idea what we were likely to find. We knew a little about the lake, but only very little. And nothing at all about the river.

The expeditions of the early sixties had been the beginning of real science in these valleys. They had gone beyond those wonderful anecdotal accounts of Scott and his men, who had talked about the light and the silence and just the utter strangeness of the fact that there was water in these frozen deserts at all. In the sixties, geologists like Ernest Angino and Kenneth Armitage and Alex Wilson were down here drilling through the ice, making temperature and bathymetric measurements, and collecting every water sample they thought they could get shipped out of McMurdo. From the outset, Vanda was like nothing that they or anyone else had ever seen. Their first thermocouple data on the water column showed that the lake actually became warmer with depth. And not just warm. Considering where they were, it was hot! Angino and Armitage recorded bottom water temperatures of twenty-five degrees Celsius in Vanda, and this was in a land where the average air temperature was twenty degrees below zero. Here was a stratified lake, but the stratification was upside down. Armitage tried to the publish the paper in one of the more august journals of limnology, but the reviewers just laughed. They told him something must have been wrong with his temperature probe. Temperature probe nothing, he had said, I felt the water. It rolled over my hands. It was hot!

But that was just the beginning. Once Angino and Armitage had gotten back to Kansas and had set to work analyzing their samples they found something even stranger. Geochemists have catalogued the Earth's surface waters from the mouth of the Hudson River to the bottom of the Cariacco Trench. And despite the infinite variety of water chemistries, there are just a few general rules that bring it all together.

First, since water is as nearly a universal solvent as anything we know, its molecules tug at the seams of matter and disintegrate it. What had been bound becomes free; what had been locked and sequestered becomes opened to the world. But the way the world is constructed, with its caging and bones, of granite and basalts and shale and limestone, ensures that only certain ions will fly from the fold in great abundance. Geochemists call these the "major ions of natural waters."And it is perhaps cause for some small rejoicing that there are only seven of them: calcium, magnesium, sodium and potassium—the cations; and bicarbonate, chloride and sulfate—the anions. Positive and negative, swarming as if possessed, from the body of the stone.

Thus the cations and anions distribute themselves in only certain combinations, so that it is possible to make generalizations. Very dilute waters, the waters of the tropics, waters in the canopy of lush Malaysian forest, contain mostly sodium and chloride. Their chemistry is determined by what is in the atmosphere, the lingering presence of sea salt recycled on its long journey back to the land. By contrast, the waters of the great rivers, waters that drain the great continental landmasses, tend to be more concentrated. They have more ions in a given volume, more ions per liter, and they have more than their share of calcium and bicarbonate. Finally, there is a third class. These are the waters that have been concentrated over time—the brines of Magadi and Trona and Lake Qinghai, and above all, the waters of the sea. Almost invariably, these brines are rich in sodium and in chloride and often in magnesium and bicarbonate. The ionic concentrations in these waters are regulated both by evaporation—the slow drawing away of one water molecule from another at the surface—and by precipitation, the joining together of ions deep within the solution to form a solid phase. When surface waters evaporate, calcium and bicarbonate join to form calcite, or calcium and sulfate precipitate as gypsum. No wonder then that saline waters, so rich in sodium and chloride, are so relatively depleted in calcium.

All, that is, except Vanda. What Angino and Armitage had found was that the deep Vanda brine—seven times saltier than seawater—was composed of mostly dissolved calcium and chloride. Not sodium and chloride, like the Great Salt Lake or the Dead Sea or a thousand other brines around the world, but calcium and chloride. It was the kind of thing you just didn't expect. In fact, finding calcium in such large quantities in a chloride brine was something akin to coming upon Chartres Cathedral atop the Acropolis. A pure anomaly.

When they wrote about it in the *University of Kansas Science Bulletin*, Angino and Armitage had some tall explaining to do. How could a body of water with this chemical composition ever form? They reached into their geochemical bag of tricks and came up with some plausible stories. Maybe an arm of the Ross Sea had once extended up into Wright Valley. When the sea retreated from below, the mountains and the water began to evaporate, the salts became more concentrated. At the cold temperatures of Wright Valley the normal sequence of mineral precipitation was altered. The last salt to remain in solution was calcium chloride. Or maybe it was like this. Maybe the salts from the stream waters that flowed into the lake had been concentrated over thousands of years. Maybe those stream waters had just the right composition to produce a calcium chloride brine. Or maybe something else; maybe volcanic springs were entering the bottom of the lake. That would explain the high temperatures. There were so many possibilities and so few constraints. Anything could have happened. Angino and Armitage had written a compendium of possible worlds. Canfield and I had wanted to know which of these worlds was real.

Canfield was extraordinary. The New Zealanders at Vanda Station called him "Hey Man," or sometimes, more formally, they called him "Hey Man, What's Happenin'." They kept a book on him, wrote down his every expression as if they were recording a strange new language. "What's up, Jack?" they asked him when he came through the door out of the bright evening sun. Father Jeff, the chaplain at McMurdo, referred to him as "your cheerful companion," or "your friend the enthusiast." Canfield worked without letup—fifteen to twenty hours were nothing for him. And it didn't matter much whether he was carrying out a tedious ten-step analysis for phosphorus, or fine-tuning a differential equation, or pulling a sled over seven miles of rough lake ice. And when he was finished he could knock back with a few Steinlagers or a shot of Beamo and joke and banter and trade stories or talk science with you all night. He seemed to enter another world when he worked on a problem, as though he were practicing Zen meditation. Sometimes you just had to shout at him, at point-blank range, "Hey, man, what's up?" to get through.

But Canfield was as puzzled as I by the origin of the lake. If the calcium chloride brine in Vanda had not come from an ancient sea (strontium isotope work had ruled that out) and if it had not come from the evaporated waters of the Onyx (a few thermodynamic calculations showed this), where had it come from? We talked about it endlessly— as we dragged our sleds to the hole, or as we sat eating dried cheese out on the lake, or while we were having hot scones and tea with the Kiwis at Vanda Station. Back in Ohio, we talked about it at the kitchen table, over coffee and preserves, with Wanda there smiling and wondering good-naturedly, "Why in the world are you doing this?" We talked about it at Ozzies, on High Street, over a pitcher of beer, where we were loud and animated and full of ourselves. I remember a young woman coming up to Canfield and saying, "You sound so involved. You must be members of the campus socialists." "No," he replied, "We're geochemists. We're talking about a lake." "Oh," she said, "I thought it was something important." Or we talked about it in the lab at Boyd Hall, where we would break from our work on the trace metals and get back to the "origin problem." I would be sketching the lake on the board and drawing arrows where the river and the small streams flowed in, and saying, "Well, maybe there really are springs, like Angino said." And Canfield would remind me, "No, that can't be true. The temperature profile is too regular."

This went on for two years. And then, by accident, I came upon an article on the brines of the Canadian Shield—the dense, deep-dwelling waters of the northern granite shield, which were about as far from Vanda as you can get. And I plotted the major ion composition of these waters on a graph that I kept on the wall in my office, a plot that contained every lake and river and pond and sea and even puddle that I had ever studied or read about. And when I displayed the points for the Canadian brines, they fell well beyond the boomerang-shaped envelope that contained most of the surface waters of the world. On the figure, they plotted high up on the left-hand side, next to only one other point—the point for Lake Vanda.

After two years, I could hardly believe it. I ran to the library, pulled down from the shelf a volume titled *Dry Valley Drilling Project*, and turned to an article by Cartwright and Harris. There it was. Cartwright and Harris had found evidence, a decade earlier, for a deep calcium chloride brine lying beneath the Vanda Basin. They argued that sometime in the past the waters from this deep source had penetrated the lake bed. Their hydrological findings had been largely ignored by geo-chemists, who were, after all, members of a different tribe. But on reading their paper in this context, in the midst of all that we knew and had speculated about, it was clear that their groundwater hypothesis had to be part of the story.

It was October, and Canfield had gone east by then. But we were still corresponding on the origin problem. I wrote him immediately with the news. I told him that the similarity between Vanda and the Canadian brines had to be more than just coincidence, and that Cartwright and Harris's deep groundwater must be the source of Vanda's unusual chemistry. Still, we knew that the river, with its bil-lions of gallons of flow each spring, must have brought, over thou-sands of years, its own treasure of ions and salts to the lake. Had the salts come from two sources?

He couldn't have had my letter for more than a week when I received a reply. It was not so much a letter as a hastily scribbled set of calculations he had done on a napkin. Among the coffee stains were a few equations and a table with two columns headed "Onyx End-member" and "Brine Endmember." He had developed an equation, a mathematical model for the lake. The model told him from which source the seven major ions had come—from deep groundwater or from the river. From the model he had calculated how many of each ion there should be in the brine and, from our analyses and from the geometry of the lake, how many of each ion there actually were. The agreement was remarkable. In a simple mathematical equation, Canfield had captured a tiny part of the world. At the bottom of the napkin he wrote, "Hey, man, what do you think?"

I finally made it to Bull Pass by midafternoon. On a ledge, I sat four thousand feet above the valley floor. Behind me the broad pass fell gradually away toward McKelvey Valley and Lake Vida. Before me it dropped precipitously to the channel of the river. All around were cavernous boulders, wind- and salt-eaten, hundreds of tons of stone and energy rafted into the mountains. Great dolerite towers, soft-hued and wind-weathered, rose to the sky. And below on the geometrical Earth the frost-heaved polygons lay rimmed by dry snow, and latticed and bound and stretched as dark carpetry to the lake. It was all hori-zontals and scallops and curvatures, parabolas of rust-colored stone and the faraway hills tan and white with their quilting.

Here, the great U and scoop of the valley was visible as nowhere else, the glacier's gouging set in stone. From the Asgards, the lateral moraines of the Vanda, Loop, and Pecten ages rose like swells on a troubled sea, then collapsed toward the river course. Promontories and headlands, dark peninsulas of basalt jutted into the white afternoon of the lake, the interfingering of ice and land, of black and white, like a winter-locked Rhode Island coast. Along the mountains were traced the faint limnings of an ancient lake, of Great Lake Vanda, as it was three thousand years ago, deep and broad and filling, the valley's quench and succor. In the distant time of Sumer and Gilgamesh, in the time of the first cities, Vanda had stood alone in the long fold of these mountains, a frozen inland sea.

I must have been sitting there for a long time. When I looked up it was snowing. Tiny crystals were clinging to the sleeve of my parka and gathering on the coarse ridges of my gloves. Up valley I could no longer see the Dais, and Vanda had blended in with the sky. In a few minutes I could barely see my hand.

All over the exposed regions of Antarctica, the rocks and boulders have been hewed and hollowed by salt and moisture. A windblown crystal of halite or gypsum or antarcticite will settle into the cracks of a granite boulder and begin its work. At first the stone barely notices. But then something begins to happen. A little moisture, a droplet of water—just a few billion billion molecules—enters the salt, and the molecules begin to lock themselves into its crystalline structure, the charges of the cations holding the water's oxygen-end with fierce attachment. The crystal swells, its lattice opens. The rock begins to flake and foliate, to turn bit by bit into concavities as though it were a wave folding over the land. At Bull Pass the salt-hollowed forms are extreme, as huge as a dwelling, as inviting as a cave. I crawled into a massive "clamshell," thousands of tons of rock at the edge of a windswept pass, and rolled against its hinge, drew my hat down over my eyes and my parka around my head, and prepared to spend the night. Outside, the snow was falling thick on the slopes. In the distance I thought I could hear the river breaking over stones. But it was only the deceptive wind.

▾ ▾ ▾

There are such dreams in daylight here! Everyone speaks of them. Long and slow-moving, unfolding like cinema, a whole lifetime running through. In the stones of Bull Pass, I dreamed of my father. It was as though he were here, a visitor for the night, up in the mountains with me. I could see the color of his face, the veins standing on his cheeks. He was a young man first, maybe thirty-five, maybe forty, the man in the hallway just entering the door, the cold air of winter work in the city clinging to him, an aura of cold around his broad shoulders, a shadow that I stood in. And snowflakes were on the black wool of his uniform, and he was wearing a white cap that froze my hand when he gave it to me. And he said nothing. Just stared out over the dry riverbed and shook his head as if this were nothing he had ever seen before. When he came again, his hair was white. He was in the train station. "You don't have to see me off," I said. "The train's so late. You should get some sleep. Take Mother home. She looks tired too." But he wouldn't leave. And we sat until three in the morning, not saying much, until it came. I saw his face through the window along the tracks disappearing into the night as we moved east. New York. London. Belfast. You could see the glow of the mills moving on the night sky like an aurora. It was the last time I saw him. Except for the dreams, when he felt so near.

NINE

The Sea

I told her I would bring her a seashell from the continent. And she said, "From Antarctica? But there can't be seashells there. The sea is frozen solid." "Yes," I said, "but I wasn't thinking about the sea. I was thinking about the valley. There are shells there. It's like the Cape," I joked. "A little colder, maybe." My mother laughed and said she would have to wait and see.

▼ ▼ ▼

WHEN I AWOKE, the sky was a deep blue. Most of the snow had already sublimed—had turned, in that most magical of physical shortcuts, directly from the solid to the vapor, the molecules of last night's condensate, bursting the bonds of substance to join the air. Cumulus clouds were fluffed all along the ridges of the Asgards just a thousand feet above the glaciers. It was late morning and I was concerned that Tim, Mike, and Dr. Yu might be worried. But I had told them that if I got pinned down by the weather, I would find myself a good rock for the night. "It might be the next afternoon before I get back to camp," I said. "So don't worry."

When the wind had stopped and the low sun was shining out of a clear sky, the cold meant nothing out here. At zero it felt balmy. Days like this had earned the Dry Valleys the reputation of being the continent's "banana belt." "So you're working the banana belt," some hapless, frozen astronomer back from a month at Pole Station would say. "Must be nice. I've heard it's a real Riviera there." Then you would see a smile flash out from beneath a black beard.

As I descended the steep slopes, airborne, jumping from one level of weathered slide-stone to the next, clouds of white and gray dust and salt rose up where I had landed and hung like contrails. It took only ten minutes to get down. "When I looked back, I saw a line of haze running zigzag up the slope. I stood and watched it settle against the backdrop of the sky until the air had once again become transparent. Direct sunlight mingled with bitter cold—a childhood memory of one's sweetest winter day. I could not return to camp just yet.

▼ ▼ ▼

Instead, I walked over toward the dry riverbed. I was to the west of Bartley Glacier. About a hundred yards from the Onyx, I came upon a gully that had been cut by a small stream that flowed out of Bull Pass. On its way to the Onyx, the stream had weathered the ground down to a stratified layer of gravel and sand. The layer was perhaps only a few inches thick and it had once been buried under more than ten feet of glacial drift. Over the millennia, the stream had worked through the drift, had swept away rock, until it had exposed the yellowish deposit below. In this deposit, more valuable to geologists than a seam of gold, lay the ruins of *Chlamys tuftsensis*, a species of pecten—an extinct clam.

In the early sixties, about the time that Angino and Armitage were puzzling over the warm waters they had drawn from the depths of Vanda, geologists were hiking these valleys in search of clues about the glaciers. In what distant past had they come? In what sequence? From which direction had they advanced—from the plateau or from the Sound? How massive were they? What landforms had they left behind? Among the more curious deposits were the fossiliferous gravels, the pecten-rich material near which I was standing.

From where had these shells come? What stories did they tell? How were we to imagine the world (this world!) of which they were a trace? There were many hypotheses. Science knew no end of conjecture, of revolutions like tiny cloudbursts breaking out every day. Perhaps the shells were transported to this location by those grand agents of dispersion, the wind and the birds who ply it. But, if that was true, wouldn't they be more randomly distributed, not cased in a single, ordered stratum? Or perhaps the shells were part of an old beach deposit. Perhaps they were laid down in shallow waters at the margin of a fjord. It is thought that once, in Vanda Time, the Wright Valley may have been an extension of the sea, like Framvaren in Norway or Sogne Fjord near Bergen, or like the whole notched coast of Greenland.

According to this argument, the glacier would have weighed upon and suppressed the land to such a degree that, after its retreat, water from McMurdo Sound could have spilled freely into Wright Valley.

There are problems with the fjord hypothesis, however, not the least of which is the fact that the rock threshold of the present valley is nearly a thousand feet above sea level. Given evidence from other glaciations, including those in Wright Valley, it seems unlikely that even the massive Wright Glacier of Vanda Time could have depressed the land to this extent. And even if the valley had been a fjord, the pecten shells, given their present location, would have had to lie in deep, quiescent water. The sands and gravels among which the calcareous shards are interspersed suggest that they were, in fact, laid down in turbulent nearshore waters—near an ancient beach. Not in the serene depths of a fjord.

But what beach? An article written in the days of speculation, in the days when the valleys were still new to science, presented an intriguing idea. What if the shells were not deposited in the valley at all? What if they were really carried in from the sea, from McMurdo Sound, by the moving ice, by a glacier pushing its way westward up the valley? Similar fossiliferous tills had been identified near Boston and near the English and Irish coasts and, better yet, on Ross Island itself, where marine fossils had been found in tills churned up from the bottom of McMurdo Sound. There is good evidence that the Sound had, on one or more occasions, been filled with ice, rising at times as high as a thousand feet above present sea level. That the immense ice sheet that once pressed through McMurdo Sound could have branched and flowed a mere fifteen miles into the valley and could have deposited its marine plunder here, below the mountains, seemed altogether plausible. It this is correct, then the shallow-water pecten that had once flourished on the floor of the sound were brought by the ice into the valley and redeposited, in a matrix of gravel, by the meltwater streams. The remains of the sea were brought to the land, and the world, rearranged, rolled on. All of this happened very long ago, in Pecten Time, 800,000 years before the present.

I sifted through the sea wreckage, through ice-fractured splints of chalk, among the sands and gravels of the stream, in search of a shell. But there were only fragments. Then, five inches below the surface, just above the layer of unweathered till, my fingers traced the rough edgings of a single valve. I caught it with my thumb and forefinger and carefully pulled it toward the surface. It was intact. I brushed it off and placed it in my parka.

When I turned toward the lake I saw a figure approaching. It was Mike. "Hey, boss," he drawled in that deep voice. You could hear it rumble on down past the Denton and Meserve, down the channel of the Onyx. "Are you okay?" I nodded and smiled broadly.

We started back to camp. The wind had picked up, blowing out of the west from the plateau. Suddenly it was bitter winter again. I raised the hood of my parka and pulled the sleeves down over my wrists. Mike was walking like the field geologist that he was. Long strides, serious, no messing around. "We had good coms with McMurdo today," he said. Walt gave us a full report on the phosphorus. I'm no chemist, but it sounded to me like one of those reagent problems. "When he put the samples into the autoclave for total phosphorus, the solutions turned blue. The phosphorus reagent changed nothing— same old color."

"Sounds like we're picking up impurities from somewhere," I said. "Did he rinse the glassware with hydrochloric?" Mike didn't know. Impurities. From where? "Which of the dozen chemicals that went into the analysis? Most of the time you could run blanks on everything until you came up with the source of the contamination. But sometimes it wasn't that easy.

Several years ago, Canfield and I had run into this difficulty with ammonia. We had collected a set of samples from Vanda, had brought them back to McMurdo, and had set up to do an ammonia analysis. Everything was ready. The samples were lined along the bench. We spiked them, expecting a nice gradation in color from left to right. But instantly, all of the flasks turned ink blue. We spent a week tinkering with the glassware, the reagents, the water supply, but we could find nothing wrong. Then it occurred to us that our lab was just down the hall from the Biolab's one toilet. Could the ammonia be drifting in like some pestilence? We set up flasks of distilled water. Some of these we exposed to the air; others we stored under a nitrogen atmosphere using ground-glass stoppers. We let them stand overnight. In the morning we rushed over from the Hotel before breakfast to see what had happened. I didn't even bother to put on my shoes. The unstoppered flasks looked hopelessly murky, as though they had been attacked by a squid. The flasks that had been sequestered for the night sparkled like a fine Riesling. The problem was solved. On our next flight to Vanda we packed up a spectrophotometer and did the whole procedure on the lake—surrounded by the purest air on Earth.

Impurity problems could drive you crazy. They stopped the project dead in its tracks, and usually they stopped it just when you were beginning to get carried away by it all. You had come here, or to some other distant lake, in search of answers, and just as you were ready to find them—bingo!—everything turned blue or brilliant green or Day-Glo orange; or they went skittering right off the instrument's scale. And then you were really sidetracked. You were on a quest that had no meaning. You had to stop looking for the murderer and search for some thug who was messing with the investigation. Worst of all, you had no idea when, or whether, you would resolve the problem. There were no time limits placed on these things. They could go on forever, and sometimes they did. Sometimes a whole project turned out to be nothing more than a whodunit, with no real science involved. Sometimes the hunt went on long enough to give you migraines or to make you wish you had chosen another career.

As we slipped along the smooth lake ice, I was thinking of how Walt must be taking this. This was his thesis project, after all; he was betting his future on it. He had only a limited time to find out how much phosphorus was entering the lakes. That was a problem with this place. You had so little time and all of it was precious. Nothing could go wrong or you found yourself strapped into the Herc with an empty notebook, waving good-bye to Mount Erebus. It was not pleasant. "Twenty-two thousand miles round trip, for what?" you said.

And so far, what had happened on this trip? Varner was already packing his bags; that was a huge hole in the project. There didn't seem to be any way of retrieving the traps, and the phosphorus technique, the technique we had practiced all summer, was not working out. What more could go wrong? Maybe the streams wouldn't flow this year. Maybe it would be too cold for the ice to melt. It felt that way.

I asked Mike how things were going back at camp. "Well," he said, "we had a great meal last night. Dr. Yu cooked up some rice and vegetables in a big pot, gave us some bottles of Tsingtao beer, and we all sat outside in the snow and watched the Dais disappear." We were passing one of the holes we had drilled, checking to see if it was still there. "That's the good news," Mike said. "The bad news is that we're out of Mogas for the jiffy drill, and we're out of bamboo. So we haven't gotten a lot done." His voice was trailing off, barely audible. Looking at the hole, which now had a thick plug of fresh ice over it, he picked up again. "It's damn frustrating to lie on this lake and look down through these holes, and know we're just inches away from the trap lines. But they might as well be back in Ohio."

Mike started walking faster. He was throwing his arms in the air, clearly angry. Two days ago he had been running up and down the moraines with a silly grin on his face and a tire pump held like a gun in his hand, yelling, "I think I got one. Those crafty Antarcticans. They flatten themselves against these rocks. They blend right into the sand. You can't see them until they come up firing. Then it's too late. They got you." Now it was different. His projects at Fryxell, Hoare, and Miers were on hold until we could fish up the traps. With Varner leaving, they might be on hold forever. He was looking at the prospect of an empty notebook, and I was thinking how we might have to do all of this again next year.

By now the tent was in view. An olive pyramid anchored to the Earth. It was undulating in the wind as though it were liquid. There were supplies scattered all around. Tim and Dr. Yu were crouched around the Coleman. Steam was rising from a large pot and floating off toward the Asgards. From the lake, I could see the flame under the pot, and for just a second I felt it as a force, as a warmth that was filling me. Fire in the stone valley.

Tim got up and held out two cups of hot chocolate. "Thought you might be needing something warm," he said. "How was the pass?" I told him that I had seen the old lake levels scratched along the mountains; and that I had found a perfect pecten shell among the gravels near the Onyx; and that I had spent the night in a snowstorm. "Sounds like you had a great hike," he said. "We were worried about you." "Just the wind up there," I said. "I think it was safe." Dr. Yu was smiling and shaking his head up and down and saying, "Goooood, verrry gooood," in a drawn-out melodious tone that sounded like the beginning of a Chinese aria. He removed his glove and extended a hand to me.

But there was more bad news. After Mike had left camp, McMurdo Station had radioed to say that Varner was out at the airfield. He was on the manifest for the afternoon flight to Christchurch.

I told them I needed some sleep. I needed some darkness, or what passes for darkness here, diminished light. I crawled through the tent flaps and onto the sleeping bag I had spread on the ground. I felt the warmth of the place. It was like home.

▼ ▼ ▼

I must have slept for a long time. The sun had swung behind the peaks of the Asgards. Huge shadows, whose darkness was shaped by the peaks of the mountains, shifted in fields across the face of the Olympus gneiss. Dr. Yu and Mike were asleep in the tent. Tim was sitting down by the lake, a bamboo pole across his knees. I went to see what he was up to.

The shadows were slowly changing their shapes as the sun wove in among the peaks. A gull was turning in slow circles high above the ice, the way a red-tailed hawk turns above Ohio. The glaciers were like poured mercury against the mountains. Tim had stood up and was holding an L-shaped rod in front of himself, at arm's length. It was made of two pieces of bamboo. He was rotating it, moving the leg along the smooth ice. "Where did you find the extra pole?" I asked, in a tone of voice that made the question more of a greeting. "It was on the other shore of the lake," he said. "There was a note from Canfield stuck in the bottom."

It was pure accident that Tim was here—something that was true of all of us, I suppose. You don't grow up saying, "Well, I've figured it out. I want to be an Antarctic scientist. I've heard there are lakes down there and I am burning to study them." One day you find yourself living in a trailer with Hatcher and Hall. They are both studying microbiology with Benoit, and Benoit, just by accident, is off to the ice. He needs someone to do a few analyses for him on these frozen lakes. How about it? And you say, Sure, why not? I could use a break from the lab, from the molten salts and the gases and the glass tubes, from being inside too much.

It was this way with Tim, more or less. Just an accident of place and circumstance. When I met him he was studying political science. He had enrolled in my course only because it was a requirement. He needed some physical science, and my course was general enough and, I suppose, harmless enough to look almost attractive.

That semester we talked about revolutions—not the kind of revolution that Tim had been accustomed to—the kind that sees a monarchy overthrown or a junta brought to power. We talked about a different kind: a revolution that topples whole worlds, that brings the towers of the universe crashing in about our heads. That semester we talked about Copernicus and how he had redrawn the cosmos, had swung the blue-green marble of Earth in a tight circle about the sun, had set it spinning and spinning twice; and how the magician Kepler, amid the savagery of endless war, had filled the sky with his serene figures and his music and his timeless laws; how Galileo, musing on objects everywhere falling and rolling and swinging about the world, had created, in his workman's way, a new order, a new method, a new science. And how Newton, blessed and grumbling, had put it all together under the mythical tree, had turned God's love into a force, an equation, a subtle fluid embracing it all. Here were real revolutions.

But as much as it engaged me in the teaching of it, I think that this was not what interested Tim. Nor was it the story of Dalton, Rutherford, and Bohr; nor the tale of Mendeleyev at his piano, replaying Schubert and all the while imagining the elements repeating themselves over and over like notes; nor the triumph of Einstein at his desk in Bern. It was a more recent tale, a tale of the Earth and how it was split and sliding and churning beneath, and how all of this motion, this motion that had thrown up the Asgards and the Olympus, and had fired the pools beneath Mount Erebus, had tossed up the mountains of Hawaii on which Keith's house sat—how all of this, in its subtle and submerged way, had made the chemistry of the sea. Revolutionary sea chemistry! It was this that had charged Tim's imagination, this science of broad horizons, of shelf breaks and rifts, of chambers smoking beneath the waves, of the whole Earth taken in at once. For someone who had never left Ohio, who had never set foot on a beach, the revolutionary science of the sea was something to think about. Once, after Canfield had finished a guest lecture in my class, he said, "Who's that guy over there? The one that had the great question." That was Tim," I said. "He's studying political science."

I was in a valley, in a desert, in the shadow of mountains, among glaciers. In fact, I might have been in Ürümqi, in the vast interior salt basin of the Tibet Plateau, the most inland place on Earth. Yet at every turn I was reminded of the sea. We are never really far from it. In Oxford, Ohio, buried just beneath your feet, beneath the familiar asphalt and pavement, beneath the water tower on the town green, and the lush suburban lawns patched together into the bean and corn and pasturage of Midwestern farms, and beneath those farms, rising on a gentle acclivity into the west, into the autumn orange sun, lie, for a thousand feet down, the sediments of a great inland sea. Ohio rests on it, on the stacked and banded mountains hidden, but only by the tiniest film of dust, below. All of this came from water—water stones just beneath the land. If this mighty assemblage of marine crinoids and brachiopods and shelled creatures of all kinds were suddenly to dissolve away, Ohio would just as suddenly drop below sea level. When Wanda and I bought our house, I made a path of these Ordovician stones, dense with the remains of a long-ago sea, holding tight to their memories.

From the arroyo where I had been earlier in the day, the sea was only a few miles away. It was not visible, but I knew exactly how the valley met it, how it bowed convexly toward the Asgards and then curved gently northeast to the Wright Lower and the Wilson Piedmont glaciers. Beyond that was the Sound, and the Ross Sea and the Pacific—the white, frozen prairie stretching on forever beneath the wind. I had the picture fixed in my mind. But even if I hadn't, I would have felt it, the way you always feel these things, feel their pull and their exhalations, as something immensely present.

But there was something more. There was the lake itself, which always seemed, in its enclosed state, circled as it was by the land, like a tiny ocean—a tiny ocean into which a single river flowed. And the problem of the lake, the conceptual problem of its origin, was like nothing so much as the problem of seawater.

"Who is the sea? "Who is that violent being, violent and ancient, who gnaws the foundations of earth? He is both one and many oceans," Borges wrote. And it is true. He is both one and many oceans—the sea of rivers and sunlight, the sea of darkness and fire. That was what had fascinated Tim, those commingled processes stretching back into the dusky beginnings of things, and how it was only very recently that we were coming to understand them. It was a revolution in the way we saw the Earth. And maybe in the way that we would someday see ourselves. Science works that way—you think you are measuring the swing of a pendulum, or the fall of an apple, or pointing a simple glass tube at the moon, or playing with the weavings of space and time, matter and energy, but you are actually remaking the world.

In the nineteenth century it appeared that the seas had come exclusively from the rivers. The Irish geologist John Joly did a wonderful calculation using just this assumption. He plunged whole-heartedly into one of the more passionate debates of the age—a name-calling, insult-trading debate that pitted eminent physicists like Lord Kelvin against such profligate, time-squandering geologists as Charles Lyell, and against the very prince of "extravagant time," Darwin himself. Kelvin had calculated, based on the time that it takes a sphere of molten rock to cool, that the Earth must be relatively young—say 60 million years. Lyell and Darwin needed more time, a billion years, maybe more, to turn mountains into dust and bring forth the creation.

Joly offered this as arbitration: Suppose that the sea was once fresh; suppose that the rivers carried their dissolved salts—sodium and chloride and calcium and bicarbonate, magnesium and sulfate and potassium—into the deep. How long would it take, given what we know about the Thames and the Seine and the Liffey and the Danube, to deliver all of the salt to the sea? As a marker he used sodium. For it was believed that once sodium had found its way to the sea, it would stay there for eternity. What could remove it? So with a few bold numbers, Joly dated the world. It would have taken the rivers a hundred million years, he said, to do their work. This was a result that hardly anyone liked. It was too short for Darwin and too long for the irascible Kelvin. But the method had a certain elegance and grandeur, nonetheless; it was like the Greeks using the shadow of the Earth to get its measure. Wrong, perhaps, but lovely.

And it *was* wrong. Joly had imagined that sodium, once in the sea, would always remain there—a voyager caught in that vast single realm for all of time. What he had not known was that, like all of the other elements, this one too traced a great curve through the world, a curve that arced back on itself like a Ptolemaic epicycle.

The sea keeps nothing, returns all. That is its secret, its identity. Sodium is no exception to this. In the continuous cycle of exchange between the oceans and the land, sodium, like a weary Ulysses bound for Ithaca, returns. It happens this way: Everywhere on the effervescent sea, bubbles are exploding from the spume of breaking waves. I read once that, at any given instant, the area of the Earth's surface that is covered in foam is as large as all of North America. Into the roiled waters of the microlayer, a bubble of air is injected. When it reaches the surface it breaks. Into the cavity that is left behind, seawater rushes in a fierce stream and then, in reaction, explodes upward as a tiny jet into the air. The jet breaks and beads, becomes droplets suspended a few inches above the wave. The airborne water evaporates and leaves behind a mote of sea salt—a grain of dust to be sent aloft. Most of this sea salt is sodium chloride. Once dissolved in water, then carried by water like a passenger in a plastic sphere, then finally freed, sodium begins its homeward flight back to the land.

The sea has other ways to ensure this exchange. Some of the sodium is buried in waters that cling to sediment particles on the ocean bottom—"sediment pore waters," geochemists call them. Some of it is taken from seawater at the mouths of great rivers, where, as Sayles and Mangelsdorf showed, clays from the continents will collect sodium on their surfaces and, in exchange, return calcium to the sea. Some—most, in fact—is lost in evaporite basins, where an arm of the sea becomes trapped by the land; here the water is slowly withdrawn through evaporation, leaving only a glistening layer behind. It was in this way that the lightless salt mines below Lake Erie, and the halite mines of Poland, were formed. In this way, in part, does the sea repay its debt. But sodium is lost from the sea and cycled back to the land in yet another way. And it was this, I think, this last process, and all that it implied about the way things work, that truly amazed Tim when he first heard of it; it was this that may have nudged him a little toward science. This last process is called seawater–basalt interaction.

There was a time when geochemists were puzzled by what they called "the magnesium problem." People knew how much magnesium the rivers were bringing to the sea each year. We will say that it was about 137 Tg (teragrams), or about 150 million tons. And they thought they knew roughly how much magnesium was being lost from seawater over the same period. The known removal processes were familiar: recycling of magnesium back to the continents in the form of sea spray accounted for a loss of 3 Tg from the sea; and incorporation of magnesium, as a kind of proxy for calcium, into the shells of marine organisms, into the biogenic calcites—the benthic forams, the echinoderms, the tubular serpulids, and the coralline red algae, accounted for another 15 Tg. It appeared that magnesium was pouring into the sea in vast quantities, and yet relatively little was being removed. But if the magnesium cycle was in balance, as everyone supposed it should be, shouldn't the inputs and outputs be nearly the same? Surely the sea could not be hoarding magnesium.

Like so many scientific problems, this one could not be solved until a revolution had occurred, until the world was viewed in an entirely new light. And a revolution in the earth sciences was under way. Geologists had mapped the ocean floor, had found mountain ranges, spiked and splitting, down the middle of the Atlantic and transiting the Pacific Basin. They had begun to read the magnetic patterns of the seafloor as though they were an ancient scroll, and the patterns had begun to tell a tale. News came from the land, too, to support the once farfetched idea of Alfred Wegener—the outrageous claim that the continents were moving, had been moving over time. ("What force could move a continent?" the critics had shouted.) By the late 1960s it was becoming clear that the Earth was a giant mobile, an intricate sculpture, a vast Christmas display seen through a department store window, wherein the red and green figures advance and turn and spin in a continuous minuet. On the new Earth, in the new geology, continents could drift, seafloors could spread, the basalts of the mantle could well up and plume into the sea as pillars of fire. It was, as the geologist Harry Hess called it, "geopoetry," and nothing less. "Who is the sea?" Borges asked. And Harry Hess had answered, "Who indeed?"

In the geology of Harry Hess, Frederick Vine, Tuzo Wilson, and the other theorists of plate tectonics, rock from the Earth's mantle mingled with the sea. Could this union, this water reacting with boiling stone, have chemical consequences as well? Geochemists began to collect basalts from the mid-ocean ridges. In the tempered confines of the laboratory, they studied how the composition of sea water changed when it was exposed to molten rock, what ions were released, what ions were taken up. By the time the data were collected and sorted and graphed, it was clear: Near the mid-ocean ridges, seawater–basalt interactions were adding calcium, silicon, and potassium to the sea; and in vast amounts, enough even to balance the geochemical books, they were removing magnesium. The magnesium problem was being solved, not at the mouths of rivers or where the winds turn the waters to foam, but among the lightless ridges and valleys of the deep. Plate tectonics, and the beautiful geochemistry that came in its wake, had rewritten myth, had created with the passion and exactitude of number and measurement a more poignant myth: that the sea had come from the rivers, the sky waters reflecting clouds, and from the fiery stone that brimmed up from below. As they would say later, in answer to Borges, the sea is both fluviogenic and volcanogenic; the sea comes from water and stone.

Seated in class that day, listening to Canfield, who was speaking with his eyes closed as though he were actually down there where the lava streams met the water, Tim asked, "But, in a way, isn't that similar to what you were saying about Vanda? In both cases you have a mixture of shallow and deep influences on the chemistry. Streams and rivers, but also something subterranean."

Canfield held his chin in his hand for a minute and pulled at his beard. Then he said, "You know, 1 hadn't thought of it that way. That's really neat." There was a kind of hush as people left the room. I think at that moment something had happened to Tim, one of those little changes had occurred that just seem to cascade and gather speed across the whole territory of your life—but when you look back you can't even see them and you forget that they were ever there. I had seen this kind of thing before, a very long time ago. I was beginning to see it again. Only more clearly now.

The wind had stopped. All I could hear was the fabric of my parka rustling against itself as I knotted the string. When I held perfectly still, there was nothing. Just the eerie rush of my blood.

I reached into my pocket and touched the shell I had found in the morning. It was part of the sea, it was part of the sea's story, part of the way the sea had become what it was, who it was. But it was also part of this valley now, a memory of the ice. And it was a part of other stories, the stories of calcium and magnesium, and maybe even sodium. These elements were all locked, for the time being, in the shell. Sometimes I thought I could feel them.

TEN

Hawaii

The window looked up toward the Koolaus. When it rained, waterfalls formed in the mountains and sometimes I could see as many as seven. In the evenings there were rainbows.

▼ ▼ ▼

I WAS VISITING THE DEPARTMENT OF OCEANOGRAPHY for only a year, so I felt fortunate to have been given the office. Before I came, it was occupied by a physical oceanographer, a theorist who worked only with differential equations. They thought it would be perfect for him. A place to contemplate, the way theorists do. He had done some important work on El Niño and had a promising career before him. But sometime in late May he decided he had had enough of mathematics and oceanography. And maybe of contemplation, too. Without much notice he left for Hilo to become a carpenter. His office was vacant. So they gave it to me.

Most mornings Keith would stop by for coffee. We would talk about the manganese nodules, or Keahole Point, or about the adsorption studies I was doing with metals on manganese oxides. Keith Chave had pretty much built this department back in the sixties, not long after he wrote his famous papers on the magnesian calcites. He was in his late fifties and he devoted most of his time now to teaching and to keeping current on geochemistry. He no longer went to sea, and he rarely came into the lab. But he read and commented on every paper that went out of the department. People wanted to know what Keith thought. And he usually told them.

Almost from the day I arrived, he got me to thinking about manganese. He had been playing around with some data from the shallow manganese crusts that lie like a shiny black pavement over the seafloor along the Hawaiian chain. The crusts had been dredged up by the research vessel *Kana Keoki* from depths of 800 to 1,500 meters. Keith had noticed that the chemistry of these crusts was very different from that of the manganese nodules, burnt disks of metal-rich sediment that litter the floor of the deep Pacific. "What's going on here?" he asked one morning, as though he were accusing me of some felonious act. "These crusts"—he was puffing on a cigarette and running a nicotine-yellow finger over a graph he had just drawn—"are enriched in some interesting elements: arsenic, lead, iron, titanium, cerium, and especially cobalt. Look at how far those points lie above the line." He traced an ellipse with his forefinger around the locations of the anomalous elements. "Why? You're a chemist. What's going on?" He stared down at the figure again and took a puff on his cigarette. Then he looked up to the Koolaus, at a ribbon of silver water that was cascading over the cliffs. "Nice view," he said. "I think Lorenz treated you right."

From then until I left, we talked about manganese nearly every day. And the more we discussed it, the more important it seemed to be. I began to think the whole watery world revolved around it, and I think Keith may have believed this, too. Keith was a genial host and justly famous for his oceanographic gatherings. He was a raconteur of the first rank, and the proud owner of one of the great aeries in the western hemisphere. On Keith's deck, you were a thousand feet above the crater of Diamond Head, looking right down into the weathered caldera of that extinct volcano. And beyond that, beyond Koko Head and the sea lanes where the tourist ships sparkled at night in black waters, for 270 degrees all around, there was nothing but ocean. "This is the way it really is," he said. "Water. Darkness." You could feel the night moving out there, far off, filled with possibility, as though it were a piece of something old and large and unfinished.

It was winter, or what passes for winter in Honolulu. Which is to say that the mangoes were no longer hanging ripe from the trees. Everyone in the department was back from some island or archipelago or some little inlet or key along the vast reaches of the Pacific Rim. We were all gathered at Keith's for the evening. Keith was on the lanai, finishing a long reminiscence with some friends about their days at the University of Chicago. He was expansive, garrulous, the way he usually got at these parties. They seemed to provide a great release for him. I think he knew how much people here cared for him.

"Three kids from the big city," he was saying. "Who would have thought we'd all end up out here in the middle of the ocean? And be liking it?"

Someone said, "I think it's the wind. After Chicago, you just need the wind."

"Well, we've got that here." Keith laughed. Then he walked over to the railing where I was standing.

"Any views like this in Ohio?" he asked, gazing off to where the sea blended into the horizon. The city of Honolulu, when you lifted your eyes, was just a tiny crescent of light, a thin electric moon. "Any views at all?"

"Sure," I said. "From the roof of Boyd Hall you can see Indiana. That's another state. It's pretty amazing."

"I'll bet it is," he said.

"And from my study I can see the old Ordovician Sea," I went on. "I'm even writing a song about it. It goes like this:

Back in the Ordovician
We did a lot of fishin'
But we never, ever, ever caught a thing.
Gradually, it was clear
That the fish were not yet here.
And we'd just have to settle for brachiopods.

"That's as far as I've gotten," I said. "It might need a little work. Another bourbon and I'll be ready to do that last important line again. Then I'll tell you what we did in the Silurian and the Devonian, if you're really lucky."

"That's okay," Keith said, "I think I can wait."

Keith just kept looking out at the sea and then he began to talk about it—its structure, its chemistry, the way he thought manganese was behaving out there beneath the dark waters. "You know," he said, "I've been thinking a lot about the oxygen minimum zone as a way of explaining our problem. Manganese and oxygen—they're linked like this." He crossed two fingers and held them up. "Just like this."

Lying at a depth of about one thousand meters and extending over vast stretches of the North Pacific, there is a layer of water, a layer hundreds of meters thick, in which the concentration of dissolved oxygen is severely depleted, maybe only a tenth of what it is at the surface where air and water meet. A large part of the eastern North Pacific contains even less dissolved oxygen, five percent of the surface value. The depth of this "oxygen minimum" is shallowest at the equator (less than four hundred meters), but it deepens to the north and west. The oxygen minimum has an area somewhat larger than Asia.

There is some question as to how this oxygen-depleted region was formed and how it is maintained. One hypothesis claims that the organic matter produced in the sunlit surface waters of the ocean is oxidized—that is, recycled back into the carbon dioxide and water from which it came. One casualty of this "aerobic respiration," of course, is the oxygen molecule. It just disappears, or almost. Another hypothesis argues that oxygen is really removed from seawater in the distant, highly productive regions to the north and east. This oxygen-poor water is then transported to the southwest along surfaces of equal density. It is that water which eventually forms the minimum zone.

Regardless of which is true, the oxygen minimum is a peculiar place. Caught in a timeless suspension between the wind-driven, sun-lit aerobic stratum of the surface and the equally aerobic stratum of Antarctic bottom water that bathes the mountains and valleys of the deep, the minimum lies in the dark silence of molecular process. It is here, in this region, that the downward flux of organic carbon drives a sequence of chemical reactions known as reductions.

*Reduction.** This word is defined by chemists as the *gain* of electrons. It's one of those terms you just have to think about, scratch your head, come back to, and say, "Couldn't they have improved on this with just a little more thought?" The logic goes like this: It is the reduction of an element's positive charge through the gain of an electron, which is negative. For example, when, in the steelmaker's art, carbon bequeaths upon the molten ore its electrons to make iron—when the red oxidic earth puddles into brilliant silver—this is reduction. In the chemist's accounting scheme, each iron atom has done nothing more than gain three electrons, but in truth a miracle has occurred. We partition our whole history according to these reactions: the Age of Iron, the Age of Steel.

---

*I should point out that chemical reduction is always accompanied by chemical oxidation, the process in which an atom, ion, or compound *loses* electrons.

In the body of the sea, the oxygen minimum is where reduction occurs. It is the realm into which sinking organic matter, packaged in the waters above, delivers the reduced carbon of once-living things.

Oceanographers who have measured the downward movement of particles have noticed a high rate of passage of certain substances through this region: organic carbon, packaged as dead and decaying cells; organic nitrogen, from the coiled proteins of minute algae; ATP, the molecule of stored energy, from a trillion dinoflagellates whose tiny bodies had set the midnight sea ablaze with cold light; and DNA, the helix of things remembered and things to come, the torch of life passed on—all of these fall in abundance, in the staid processions of the sea. And it is this flux, this odd funereal procession of reduced carbon, that gives the oxygen minimum zone its curious chemistry. One detects there, for example, high concentrations of manganese and chromium and cobalt and iodine and nitrogen—all these in their reduced form, all more enriched in electrons than they would have been anywhere else except in the sediments. Given that these waters—so rich in organic carbon and in metals of many kinds, and so poor in dissolved oxygen—given that these bathed the dark sediments and sea-mounts of the Hawaiian chain, it seemed to Keith that this was the explanation for the chemistry of the shallow crusts.

"Think about it," he said. And he raised his two fingers again into the Hawaiian night. There was a gentle land breeze coming out of the mountains, and it carried the sweet smell of plumeria with it off toward the sea. "There's all this oxygen-poor water down there," he said, pointing over the cone of Diamond Head. "And where there's little oxygen there's usually lots of manganese—at least in the oceans that's true. Out there oxygen and manganese are poised like two kids on a seesaw. One goes down, the other goes up."

Keith had just said good night to a colleague. He was crouched down, trying to light a cigarette in the breeze. "When he stood up, he asked, "How does it work with your lakes down there?" He turned a little to the southwest, and raised his eyes beyond the towers of Honolulu. It was as though he could see the Dry Valleys beyond the horizon.

"Well," I said, "I'm not sure. In Vanda, something seems to be removing the metals from the oxygen-rich shallow waters and releasing them to the oxygen-poor deep waters. We're just not sure what it is yet."

"Sounds a lot like what's going on out there," he said, "in the oxy-gen minimum zone. Sounds to me like manganese is involved. Have you thought about using traps?"

"No," I said, "I haven't."

"Give it a try the next time you go down. Can't hurt. In some ways, you have yourself a little ocean there. It can tell us a lot of things; it can tell us about manganese. And you don't need ships. As for me, *this* ocean right here is just fine." He gave the Pacific a little salute with his glass. Then he said, "I'll leave all that cold down there to you."

It was late. I was only half hearing what Keith had to say. My daughters came out on the lanai, dressed in their pink and white mumuus. They were rubbing sleep from their eyes. Katy said, "Daddy, can I spin the whirligig before we go home? I want to see how far it goes." She held her hands over the railing and rubbed the stick between her palms. The small stick turned as it left her fingers, turned like a maple seed, rose for a second, and then drifted down the lush terraces of the mountain and off into the night. I watched it as it fell. But I was not really seeing it. I was seeing the Huey. I was seeing it move against the massive sandstones of the Wright Valley.

ELEVEN

An Invention

Mike said he had once read of a dream Charlemagne had shortly before his death. In the beautiful church at Aix-la-Chapelle, which he had adorned with silver and gold, with the perfect marble of Rome and Ravenna, Charlemagne fell asleep. It was after nightfall. When he awoke, he reported to Einhard that he had seen a sight so beautiful that he took it as portent. But of what? He had seen a great wall of snow. It was blue and jade green. There was a cold and brilliant fire on its face. From high above, he could see that a waterfall had formed. Like something he remembered in a distant land, in the jagged mountains of his mapped kingdom. But this water was pure silver. It sounded like spring in the sweetest years of his conquest, the years of his youth. It was cold beyond his believing, he told Einhard. His hand shivered and he drew it back. The water raced between mountains; where the mountains bent, it followed them, washing high on the naked rock. The sounds of its roaring were everywhere. They filled the valley so that he drew back in fear. The sun was bitter cold. He had never felt such cold, seen such emptiness. His lands were not like this, from Galicia to the Vistula, not like this. And the waters kept coming. Then they froze. It happened all at once, he told Einhard. Froze into a glittering sea. The long night came, night without end, and you could see the round moon cast its shadows flat onto the frozen waters that were still as death. But so beautiful. Could he bring this back, he wondered, anchor it beyond the gold rail, the glowing lamps, the sweet incense?

▼ ▼ ▼

WE TOOK TWO SLEDS. Into each we threw a trap catcher and a bunch of plastic caps so that we could stopper the traps if we were lucky enough to retrieve them. Dr. Yu and I headed off toward the deep hole; Mike and Tim headed east toward the center of the lake. We had the sled tethered to a long rope that opened up into a skinny isosceles triangle—the sled at one corner, Dr. Yu and I at the other two. The sled pulled easily and I held the rope near my shoul-

der with just one hand. We were moving toward a huge glacial erratic that lay among much finer debris on the lower slopes of the
Olympus Range. The boulder was a landmark. The line between it
and our Scott tent ran through the point on the ice where we had
placed our most important trap. This was the same boulder that
Canfield had mentioned in his note, the stone behind which he had
secured the bamboo pole. "Thought you might need this, guys," he
had written in that casual voice. It was pure Canfield.

As we approached the site, I began to feel apprehensive. The traps
had been suspended for twelve months. In that much time, anything
can go wrong. Even here, where we knew no one had been. We were
certain, and yet there was the question: What if something had been
here? What if the rope had been severed, been eaten through? In the
distance I was beginning to make out the rock cairn we had left as a
marker. As I drew closer, I could see that a curious separation had
occurred. The dark basalts, warmed by the sun, had melted deep into
the lake ice. They were looking up at me as if from a crystal sarcophagus. The white granites had hardly melted through at all. They lay in
smooth hollows on the surface of the ice. I was not expecting this. But
I should have been.

Dr. Yu stood there with his hands on his hips. He was turning in a
slow circle, looking over the valley from beneath his parka hood. "It is
like Qaidam Basin," he said in his clipped English, which was a kind
of poetry. "In Qaidam," he said, "No birds fly in sky. No green grass
on land. Stones run before wind. Mountains are knives. I must climb
icy mountain, go down fiery sea. People say this of Qaidam, of salt
lake. What say of this place?" He was facing down valley, looking over
the length of the lake, down toward the dry Onyx. Tim and Mike were
still moving across the ice, although I could no longer hear their sled.
But I was not thinking of the valley. I was not thinking of China. I only
wanted the traps back.

We had laid the traps where we had to, up and down the lake: near
the Onyx, about half a mile from the debouchment; by the peninsula,
where the lake constricts before it opens into its major basin; and a
mile or so west of the peninsula. In the deep lake, we suspended three
sets of traps from a single line. The first hung at forty-eight meters,
just at the top of the calcium chloride brine layer; the second was at
sixty meters, where the oxygen-rich region met the oxygen-starved
region of the lake; the third was at sixty-five meters, deep within the

sulfide brine. We placed the traps where the lake's long history fore-ordained.

It was twelve hundred years ago, near the close of the first Christian millennium, while the young Charlemagne was uniting Europe, that an epochal event occurred in the Wright Valley. At that time, in the years of the Crusades, Vanda had evaporated to become little more than a salt flat, a glistening whiteness in the center of which lay a shallow circle of water. Then the change came, and it came rap-idly. The climate warmed and the Lower Wright Glacier began to dis-charge water, not the springtime trickle that had gone before, but whole sheets of water. The lake level rose. The fresh waters from the glacier overlaid the dense brine. There was little or no mixing, only diffusion, the slow transfer of matter. Now it was as though there were two lakes: one large and made of fresh water that had rolled suddenly down the Onyx like a deluge; the other a dense brine, the sweating evaporite remains of the ancient lake.

It was this single medieval event that had set the structure of Lake Vanda, that had ensured that the upper waters would be light and clear and oxygen-filled, and that the depths would be heavy and dark and rife with decay. Knowing these things, we had set the traps.

As I lowered the new trap catcher hand over hand, the bottom folded at the hinge, twelve feet of bamboo slowly disappeared through the opaque slurry that clogged the hole. I was holding my breath. Then I felt the horizontal arm fall open when it cleared the bottom of the ice. I lay on my stomach, spread my legs, wrapped my gloves around the pole, and slowly turned it, first clockwise, then counter-clockwise, in small, gentle arcs, fishing for the line that I knew was hanging parallel, just a foot away. After a few minutes I thought I touched it. I began to twist the pole. It was resisting, as though some-thing were being wound on it, as though something were coming in. I could hear the bamboo creak below the ice and I thought it might splinter. I bent my head around to look at Dr. Yu. "I think I have something," I said. "But it feels like it's going to snap. It's very heavy."

He lay on the ice across from me, his head only a few inches from mine. He took hold of the pole just above where my hands were placed. We both began to turn the bamboo. The sounds were becom-ing louder, like the creaking of a wooden ship at the docks. I could hear nothing else. We were looking at each other, talking with our eyes: I imagined Dr. Yu thinking, *Maybe a little more, just a little, it's okay; bamboo, bamboo strong.* Then we were looking at the slurry and at the pole turning through it. There was a cloud of fog over the hole from my breath and his. I could feel the weight of the line in my fingers. It seemed too much. I wanted to just hold it there in place, for a few seconds, not risk anything. We were touching the traps. The particles were down there at the end of the line. I wanted to pretend we already had them.

Sometimes the geochemists made it sound like a riddle. "Why isn't the sea a copper blue?" they would ask. Not the blue of the sky, but the deep blue of a copper solution. After all, the rivers have been bringing copper to the sea from the beginning of time. And yet the seas are not full. Why?

To convince yourself of this truth, you could do the calculation on the back of an envelope. Each year the rivers pour into the oceans about seven thousand tons of copper. Over the last sixty million years, just since the demise of dinosaurs, the rivers have excavated and deposited whole mountain ranges of copper. But there is virtually none in the sea—a few nanomoles per liter, a few million tons in all.

Where has it gone? Why are the seas not absolutely deadly with copper? In just the last two decades, the geochemists have found the answer: particles.

It was one of those things you knew and didn't know. Science was full of them. Democritus knew there were atoms, little flicks of cold that pricked your skin, that made your nostrils flare. And Lucretius knew this too. You can look at *De Rerum Natura* and hear the wind as it shakes the mountains and rolls the seas and wracks the vessels that go there. And in Lucretius you think you "see" the wind, not as something insubstantial but as a maelstrom of tiny bodies, of corporeal beings, of atoms. And so on up through history, through Boyle and Newton and Descartes. But did they really know, know in the same keen predictive sense that Dalton knew, or that Rutherford and Bohr and Schroedinger knew? Or was *atom* just a convenient name you gave to things, to causes you could not fathom, to whatever invisible mystery it was that could hollow a rock or float a tree or sweep the sky clean of clouds? The Greeks knew that light was particulate, but did they really *know?* Did anyone really know about light until Newton and Young, or perhaps until Einstein?

So it is sometimes said that the Danish oceanographer Forsch-hammer, in one of the most eloquent and condensed passages about the chemistry of the sea, knew about particles. For what Forsch-hammer saw was that the amounts of the elements found in the seas were not so much dependent on the rivers that poured them in, but on what happened to them once they arrived. Applied to the problem of copper, this seems to suggest that despite the burnished mountains of metal that come, dissolved, to the sea each year, we have no reason to expect the oceans to be awash in it. What is really important is what the sea is doing. And the sea is not passive. The sea, here, is Borges's sea, "violent and ancient, who gnaws the foundations of earth." It is the site of chemical and organochemical action, the place in which elements are rendered, in various ways and at various rates, insoluble. Where ions are transformed into particles. Where particles sink and are carried away. Where things, for a time, are lost.

In the oceanography of Karl Turekian and Wallace Broecker and Edward Goldberg, the shadowy particles of Forschhammer have been given flesh and identity. So abundant are they in the rivers and oceans of the world, and so varied and pervasive in their influence—these clays, these oxides of iron and manganese, these cool flakes and platelets of calcite and cells of sinking plankton, and these circling desert dusts—that Turekian has referred to them as agents in a "great particle conspiracy." It is this "conspiracy," this great passage of grain and spore and flocculi, this pulverant Earth-wide storm, that has in great measure removed metals from the sea. It is this simple process that has cleansed the sea of its most toxic bodies, leaving behind only traces of copper and zinc and cadmium and lead.

What particles exactly? There were thousands of them. Which ones were the most effective "scavengers"—the cells of sinking organisms, the surfaces of calcite, the microscopic umber of the manganese and iron oxides? And did what happened in the sea happen also in lakes? And to what extent, and how fast? As Keith had said, "the proof is in the particles." I couldn't wait to see what particles our traps held.

The sounds coming from below had gotten sharper. The cold bamboo, as it strained, made noises that rang out like shots. We were still winding the rope around the pole, inching the trap line up. Dr. Yu and I got to our feet. We bent over the bamboo that stood a foot or so above the surface of the ice. I was breathing hard, knowing that it was time to bring it up, but not wanting to, afraid that the leg of the L would break off. We were standing in a cloud of condensate from our own breath, delaying what we had to do.

I began to raise the pole. I felt the leg touch the bottom of the ice sheet. I could almost sense the texture of the ice, its smoothness, twelve feet below. It was as though the organic fibers of the bamboo were feeding into the circuitry of my own arm. It was as though my fingers were tracing along the underside of the sheet.

The hole was only ten inches in diameter—the width of the drill bit. The leg of the trap catcher was two feet long, long enough to allow us to snag the line. Tim had designed a hinge that would allow the leg of the L to straighten when it hit the ice—on the way up it would become an I. That way it could be pulled through the hole. If everything worked.

It resisted. Maybe we had wound the rope around the hinge. I pulled a little harder, but it still wouldn't budge. The leg was down. It would not come through the hole. I got on my knees again. I removed my gloves and began to scoop out the netting of ice crystals and slush that lay in the hole, that obscured what was happening below. As I pulled the ice from the surface, new rime from the drilling floated up. Five minutes, ten minutes, I don't know how long it took. It seemed I would never clear the hole. Then the last crystal bobbed to the surface. I lifted it with my thumb and forefinger and put it on the ice.

There was blue light coming off the hole. Its sides were gently corrugated. You could see how the drill had cut. The water was so clear it seemed invisible. I could see to the bottom of the ice sheet and below. Far below.

The rope had knotted around the hinge, but through the sparkling shaft of water it appeared that there was room for some play. We began to lift, slowly, so as not to ripple the surface. My head was down in the hole, as close to the waterline as I could get. I was almost breathing the water of the lake. The leg of the trap catcher began to bend. It was creaking. I could count the degrees as the angle opened up. Dr. Yu was above me. He was whispering as he pulled, as though he were repeating a mantra: "Bamboo strong. No worry. Bamboo very strong. Very, very strong." He repeated it over and over. "I know this," he said. "I know this."

It had opened to 150 or maybe 160 degrees when it cracked. I saw the leg detach from the long pole. I saw it wobble a bit and begin to sink under the weight of the traps. The rope was uncoiling, falling away. I felt my heart stop. Dr. Yu said "Oooooooh," his voice sinking with the traps.

Then I felt a tug on the line. The rope had thrown a coil over the catcher. Suddenly nothing was moving. There was no sound. The bamboo leg hung limp in the water. The traps were attached.

We were both on our feet now. We were bringing the pole up through the water. It was high above our heads, a line of bamboo against the mountains and the sky. As it rose, there was a length of rope and then the leg and then the hook and the coil wound around it. I grabbed the rope with both hands, held it tightly. I could feel the weight of the traps. For the first time in a year I knew we would get them back. Dr. Yu was standing there, grinning, saying "Goooood, goooood, very gooooooooood." It sounded to me as if he was singing.

"You will never believe me," Pablo Neruda said, "but it sings, the salt sings, the hide of the salt plains, it sings through a mouth smothered by earth." It sings, but we cannot hear. A voice beneath things, but Earth-wide, and everywhere, not just in the hide of the salt plains or in the rounded grapeskin of water. The comings and goings, the small visitations, the nanosecond or millennial lingerings, the departures—there is music in all of these. Not the music of a few distant spheres, but the music of a million cycles, like hooped bracelets, fine-spun silver twirling and whispering.

In this valley and in this lake and in the sediments beneath, these cycles turned in miniature, in a space that was comprehensible, but that was tied nonetheless into the larger space of the world. The river came in the springtime. It was sound and it was light, but it was also the head-over-heels tumbling of each water molecule, the combined energies of those molecules, their separated charges like torch fires, burning at the tips. What sound did the loosening of cobalt make, the adsorbed ion wavering a little like a minnow at the surface of a rock, then heading off downstream? How long did it stay in the lake after it had glided there on the current, after it had moved faceup, eyeing the blue Antarctic sky through the prism of ice? Maybe a year, maybe five. Not as long as sodium or chloride; longer certainly than iron. Then what? Perhaps an encounter with the surface of clay, glazed with a few atom-thicknesses of manganese oxide. Then capture. The cobalt transferred from water to stone, perhaps oxidized even, an electron transferred in the wink of an eye—now you see it, now it's gone, over there!—from the cobalt to the manganese. The stone sinks, first swiftly through fresh water, then more slowly through salt, the cobalt all the while clinging, being basketed and woven in like Moses by the manganese.

And this is the way it goes. A downward journey of a few weeks. Transit out of light. Transit into darkness. Transit into the deep. And in the oxygen-poor waters the manganese is reduced, falls away, unravels like thread. The atom of cobalt is free again, waterbound. The dazzling little motions of the water molecules, like tiny boomerangs whizzing about it, coming close to its charge, then retreating, are familiar and welcome. So it stays. Perhaps a year. Then another encounter: Something that was once living, a few cells still clinging together drift by. To the cobalt it is as though the roots and branches of a great elm were being dragged by in a flood. The branches reach out, enfold it: chelation. It is on its way to the sediments. Possibly to a small eternity there. Until the next ice sheet comes. But even buried you can hear it, you can hear the cobalt. Like the salt plains, you can hear it sing.

This was only a hypothesis, of course. Possibly it was extravagantly wishful thinking. Who knew whether the cobalt's fate was linked to manganese? I wanted it to be, but the proof was in the particles. I wanted to imagine it drawn from the water to the oxide surface and then released. I wanted to imagine the sound, the susurrus of that exchange, repeated over and over and over in every lake and ocean and river on Earth, and for every element, modulated and toned and hallowed and joined in a single lifting chant. I wanted to speak the name of every element, every zinc and copper and mercury and lead, every iron and manganese and every compound of these and of hydrogen and oxygen and carbon and millions and millions more, intertwined and twirling on the wrists of the world.

The hypothesis was not without foundation, however. One year, Canfield and I had sampled for manganese and the trace metals. We did the manganese right there on the lake, on our knees on the ice, as soon as the samples came up. As soon as we got the points off the spectrophotometer, we plotted them on a sheet of graph paper. At the same time we plotted a set of oxygen values. It was remarkable. The data sets were mirror images. Where the oxygen was high, the manganese was low. Where the oxygen was low, the manganese was high.

The little decrease in oxygen below fifty meters was matched by an increase in manganese at the same depth. And when the oxygen disappeared at sixty meters, the manganese rose to its highest value.

When I saw the graph, I thought of Keith's party and of what he had said about manganese and oxygen, how they were like kids on a seesaw, how when one went up the other went down, how in nature they seemed poised like yin and yang. I thought of oxidation and reduction, of the oxygen minimum zone above the crusts of the Hawaiian Ridge; about the hypolimnion of Acton Lake.

At fifty-five meters, manganese was being reduced, it was gaining electrons from all of the decaying carbon down there; you could see that in the profile. It was as clear as anything. I began to construct a little story about the manganese. I didn't know whether it was true or not, but I knew eventually we could test it and find out. That was the way science worked. You wrote a story. It was pure imagination bounded by a few, usually weak, constraints. Then you tested it, saw whether the world out there could really abide your notions of what was so. Usually it could not. So you tried again and again until you got it. Until you had something that might actually be so.

It appeared for all the world that manganese—probably in the form of solid manganese oxide, probably clinging to the flat surfaces of clay particles that had been weathered by the Onyx—was sinking into the deep lake and was dissolving away there into manganese ions. The solid, with a sigh and an exhalation, was becoming mere charge in the reducing, electron-rich waters below fifty meters. But if that was so, shouldn't there be a great release of other things as the oxide fell apart and crumbled in upon itself? Shouldn't there be a great release of cobalt, for example, or nickel or other metals that might be riding upon the oxide surface? Shouldn't the profiles for these tell a similar story?

I remember how I had looked forward to the metal analyses. No sooner had the samples arrived back in Ohio, in their sturdy wooden boxes, than I was looking at them. When you study metals, you become obsessed with purity. You can't escape. I worried about every stray breeze, about whether it carried with it a scintilla of lead or cobalt. I worried overtime about the reagents. Every working hour—which is to say every hour I was awake. Was the Freon TF as clean as it could be? What about the DDDC and the APDC and the nitric acid? And the bottles in which the samples had been collected, and the separatory funnels of expensive Teflon? I began to dream of metals.

They were everywhere, truckloads of them, and I was counting every atom. Just another of my counting dreams!

After treating them and going through all of the preparatory steps, I had only a tiny extract from each sample. I took the extracts and lined them along the bench in Boyd Hall. The windows let in the winter light from outside; the trees were bare, snow had fallen in the woods. I switched on the instrument, dialed in the wavelength for cobalt, turned on the argon tank and the cooling water, and programmed in the temperatures for the graphite furnace. Into the cups of the automatic sampler I put a few milliliters of extract from each depth, in exactly the same order in which we had collected them. For a second I imagined I was in the field, facing the lake.

Then I set the instrument going. The little arm of the autosampler began to twist. Then it came down and took a few microliters from the cup, barely a drop. It halted a second before it rose again and moved toward the graphite tube. It looked like the pitching machine my father had used for throwing batting practice. Just like that. When it reached the tube, it hung there, deposited its precious droplet, then came back and rested. The furnace kicked in. The temperature of the tube climbed to 110, stopped; climbed to 250, stopped; then shot instantly to 2,300 degrees. You could hear the controller work. There was a surge of current through the tube. Then a burst of white steam, as though a tiny volcano had just erupted right there in the lab. A trace of light shot across the face of the instrument. A number appeared on the screen. Then it appeared on the printer. There were little clicks as the paper moved into position for the next reading.

It was working, but I couldn't watch. I was too nervous. I went back to my office and let the instrument tack downward into the depths of the lake: five meters, ten meters, fifteen meters; it was moving slowly, doing triplicates on each sample. I had the instrument interfaced with the computer. After each analysis the point was placed on a graph whose vertical axis represented depth and whose horizontal axis was the concentration of cobalt. I waited. I thought about what Keith had said at the party. About how manganese might be controlling everything in the oxygen minimum zone of the Pacific. How oxygen and manganese were linked, "like this," as Keith had said. And how the metals might be linked to manganese.

When I finally summoned the courage to go into the lab, the analysis was complete. On the blue computer screen there was a curve connecting the points, from the ice surface down through the water column to the sediments. The whole journey, a whole year, was laid out there on the screen in a single trace. It was frightening to look at. I wanted to shield my eyes.

But there it was. Point for point, the curve for the cobalt analysis matched the curve for manganese. Matched it to a T. Where the dissolved maganese was low, so too was the cobalt. Where the manganese rose, in response to the disappearance of oxygen, so too did the cobalt. Even the fine structure between fifty and sixty meters matched up. It was perfect. The story was beginning to write itself.

It was night when I left for home. The woods behind Boyd Hall were filling up with snow. The drifts came nearly to my knees. At times as I walked I thought I was floating.

▼ ▼ ▼

But as Keith had said that evening, overlooking the flat darkness of the Pacific, the proof is in the particles. If the manganese oxides were really transporting cobalt and lead and the other metals, as the overlapping profiles in Vanda had suggested, then we should see it in the particles, in their composition, in the way the metals were distributed among the complex tangle of plankton cells and minerals and bits of clay and organic ooze. Each particle was, indeed, a universe, a world more extraordinary than even Blake had imagined. On every particle there was generation and decay, the comings and goings of ions from a million knotted surfaces; breathing and exhalation conjoined on a sphere no larger than a pinhead. How many angels could dance on the skin of a sinking particle? Maybe a million. Maybe more. And how many manganese atoms and how much cobalt?

So we had put the traps in. We had left them there suspended. They had seen the coming of winter. They had seen the sun extinguished. They had seen the moon rise over Linneaus Terrace and cast shadows on the mountains. They had heard the fierce Antarctic wind, the wind of the early explorers, the wind of Scott's death. They had seen the lake scoured with sand. They had heard the ice crack like a bullwhip.

Before I saw anything else, I saw the rope. It had gone in white. But in the lake it had turned brown and dark green. Filaments of algae clung to it and wove about it. It had the hoary look of things brought up from the sea.

I had not been expecting this. Vanda was known for its ultra-oligotrophy: biologically, it was among the least productive lakes in the world. In all the water I had drawn from it, all the thousands of liters, I had not seen a single rotifer. Nothing. Even under the microscope, I might as well have been looking at a drop of liquid mercury. Thoreau had said that Walden was "not fertile in fish." Vanda was fertile in nothing. It would have made the waters of Concord look fecund and teeming by comparison.

I had done some rough calculations, before we put them in, on just how much sediment we might expect to find in the traps. There was very little to go on—a single estimate of sedimentation rate that Alex Wilson had made several decades ago. Taking into account the diameter of the traps, I calculated that there should be about a hundred milligrams. Mostly sand, I thought. Sand that had worked its way through the ice cover. But my estimate could be off by a factor of ten either way. Maybe out here, so far from the river, we wouldn't find anything. Maybe like the rotifers, there wouldn't be a single particle.

The first trap, the one we had suspended at forty-eight meters, brimmed with crystalline water. Nothing broke its transparent perfection. I raised it like a dry martini to the afternoon sun. There was nothing to scatter the light. It moved in unbroken rays through the trap chambers. Had we come all this way, I wondered, waited so long, thought so much about the particles, about the rain of manganese oxides cleansing the lake, only to find this?

"Empty," Dr. Yu was saying. "Empty. Only water." He had removed his balaclava and was scratching his head, wiping his forehead.

I put the traps on the ice, stood the chambers upright. I reached into my parka and took out some orange plastic caps. I placed these over the top of the cylinders. Then I lifted the whole assembly over my head and began to shake it. I don't know what I was expecting. Maybe that the water would somehow transform itself into earth, the way the Greeks thought it did at the mouth of the Hellespont. Maybe that it would become air. Maybe that it would burn. Water could do anything. You just had to believe.

Dr. Yu was saying, "Ohhhhhh, careful. No break." I shook the tubes for a full minute. A dry snow had begun to fall. The flakes were large. They clung to my skin without melting. I put the tubes down on the ice, lay on my stomach, and held them in front of my face. I couldn't tell what I was seeing. Something was drifting down inside. Maybe it was just the snow. Maybe the water *was* turning to earth.

Gradually I began to discern what was happening. There were tiny bits of clay suspended in the water. It was faint, but you could see it. It was settling, but very, very slowly, the way small particles settle. Stokes's beautiful law. It was the dust of the lake, the dust of the river, the dust of the valley. We had waited a year for it. I let out a whooop.

We were on the lake all evening and well into the morning. It continued to snow out of a gray sky. The whole valley was enveloped in cloud. The mountains folded in around us. A wan light spread up toward the Sound. The glaciers seemed to be floating again. The parkas whispered against themselves as we moved.

We repaired the trap catcher, and each time it held. It folded tight against itself and went down the hole. It opened and groped its way through the lake. It touched the line, gathered it in, wound the traps to itself. Then it released at the ice edge and straightened again, just the way Tim and Dr. Yu had designed it. Bamboo could do anything.

We covered the whole length of the lake, all the way down valley to where the Onyx would soon begin to flow in. We turned the sleds to shore and pulled them along the smooth annual ice, five miles back to camp. You could hear the water sloshing in the traps as we moved.

Just where the ice meets the shore, you could see a thin band of open water. It was no more than a few inches wide. Dr. Yu was pointing toward it and saying, "Ahhhhh, water! In Qaidam Basin, we say maybe spring. Maybe not long."

TWELVE

Whiteout

On the occasion of Planck's sixtieth birthday, Einstein began his eulogy for the great quantum theorist by talking of the motives that lead us to science and art. He spoke of the need to escape everyday life with its dreariness, its changing desires. To move beyond grocery lists and mounds of papers, the bills coming due. "A finely tempered nature," he said, "longs to escape from personal life into the world of objective perception and thought; this desire may be compared to the townsman's irresistible longing to escape from his noisy, cramped surrounding into the high mountains, where the eye ranges freely through the still, pure air and fondly traces out the restful contours apparently built for eternity."

Anyone who has done science has felt this, has waited, almost palpitantly, for this moment. But before the mountain there is the valley; before the vision there are the prayers and the fasts and the rituals; before the summit there is the long climb over rock and scree and fallen ice; and there are the crushing doubts that you will ever arrive, that you will ever be blessed enough.

▼ ▼ ▼

A S SOON AS WE GOT BACK to the tents, we set up the nitrogen tank. We connected a Tygon tube to the inlet of the filtration device. I shook the traps to resuspend the sediments, then I poured the mixture into the plastic cylinder that I had perched on top of a wooden box. The idea was to pass the solution, under clean nitrogen pressure, through the fine filter. The particles—the clays and oxides, the algae and bacteria, the whole peaceable kingdom of inorganic and organic substances—were captured in the microscopic weave of the filter, while all the chloride and calcium ions and the molecules of the lake went right on through.

I filtered the sample from the trap nearest the Onyx. I turned up the pressure on the nitrogen tank so that the water would flow faster.

But there was so much sediment in the trap that the process became a drop-by-drop extrusion. Toward the end you could see the drops swell and detach from the underside of the filter as though they were ripening fruits. You could count them as they fell. It took two hours.

I worked with my gloves on, with the hood of my parka raised. In the west there were low-lying clouds, but the snow that had fallen all night had stopped. The sides of the Dais were dusted white. Above the plateau there were small patches of blue sky, and through the clouds there was a golden light spreading over the upper slopes of the Asgards. The valley floor appeared undulant and broken, a chiaroscuro quilt spreading toward the thin peninsulas of the lake. A fine line of open water tracked the shore. In its surface you could see the highest peaks of the mountains. All of this was around me when I glanced up, when I remembered to stop looking at the tiny patch of clay that was coming into view on the filter.

I thought of the rutted clay as a world. I thought of myself as an explorer of that world. I imagined continents of calcite, continents of iron oxides and manganese oxides and organic oozes and crystalline silicates. And I thought of myself on these continents, boots and pickaxe and knapsack, crossing the ordered ionic landscapes of the oxides, wallowing in the molecular carbonaceous muck of the organics. As I scanned the horizon I saw the rounded shapes of cobalt and nickel and copper and lead like huge erratics dropped there under the sky. And I saw myself counting them, counting atoms as though I were Avogadro, and saying, "So this is where the cobalt is, this is where it comes to rest. On these other lands, it establishes only beachheads and meager colonies, but this is home." Perhaps.

But for the time being I could only look and wonder and guess. I would know nothing until we had done the analyses. Until we had performed the rituals, spoken the prayers.

There is too much suspense in this work. Not the kind of suspense you get from a mystery—though there are mysteries, too—but the kind you get from waiting, like a child on Christmas eve. You get the traps, you peer into them, and you can't stop yourself from speculating: "These clays, why do they appear so light? I thought they would be darker. Is that a calcite crystal in the corner? Maybe not. Is there any manganese in there? There seems to be something over by the calcite, a faint film. Maybe I'm just seeing things. And why isn't there any sand? The sediments are mostly sand. How do you get a sandy sediment from this clay?

You look and you speculate, but it's useless. There's nothing you can do to answer these questions. Not here. Not in the field. To get the answers you must wait. You must be patient. You must be willing to do certain tedious things and do them in just the proper order. Some people call this "cookbook science." They dismiss it as rote exercise, something beneath their dignity. But it is necessary. Without the gathering and cleaning, the paring, seasoning, and tasting, there can be no meal. Without these common rituals, there can be no banquet.

And then there are the samples. What if the samples didn't make it back to Ohio? Without the samples there is nothing; without them there will never be the possibility of knowing. If you were like Alfred Russel Wallace, you could lose a lifetime in your samples. What if you were that unlucky—to have crossed a hundred thousand square miles of rain forest; to have gathered every red-eyed newt and carnivorous plant and placed them lovingly like pressed flowers into vials and casings; to have gathered the whole colorful creation into the hold of a ship; and then to have set sail with it toward dreams of discovery, homeward to the lab; and then . . . and then to have had it all go up in flames? No one could ever be that unlucky again, could he? But you worried anyway. You prayed to the patron saint of deep scientific misfortune so that he might spare you these afflictions. You prayed to Wallace.

I remember one of Canfield's geochemical excursions. He had packed up the old truck, with baling wire and tape and rope holding the doors, and set off from New Haven. He had a few hundred plastic bottles rolling around behind the cab, and a map of the country folded on the seat. He followed blue highways along the banks of rivers. All along the way he sampled: the Mississippi at New Orleans and Vicksburg; the Red River near Alexandria; the Arkansas at Fort Smith; the shallow Platte wherever the road touched it; the Missouri around Pierre and Bismarck; the Snake and the Columbia. When he came to the Pacific, he had water that had drained every conceivable bedrock and sediment, every geochemical province. He was ready to study iron—how it moves, in what mineral forms, in what oxidation states, at what concentrations. But before he turned east, somewhere near the Oregon coast, he went down to the shore. On his return, as he crested the last hill, he could see that the truck was gone.

A whole continent of information gone with the truck into the cool Oregon twilight. He spent his nights in poorly lit motels along the coast, hung out in local bars, asked lots of questions. After two weeks he took the Greyhound back east. The following spring he bought another truck, no better than the first, and headed west. He collected the samples all over again.

Sometimes the patron saint of scientific misfortune just doesn't hear your prayers. But sometimes he does. One year we collected the Onyx River, hiked the channel all the way from the glacier to the lake. We had the samples choppered back to McMurdo, crated up, and mailed to Ohio. On the box, under my name, was lettered the instruction DO NOT FREEZE.

We waited until mid-February. I was still living in dream time, in the time that moves so slowly after these trips, when nothing is wrong. All the equipment had gotten back safely: the Kemmerer bottle, the pump, the gravity corer, all of the meters. I had gotten everything repaired and cleaned at the shop. Had put them away for the next season. The samples from Vanda were back. But there was nothing from the Onyx.

I started calling around: the freight company headquarters, the regional offices, the local office. No one knew anything. Then the blizzard came, and the big freeze of the decade fell on the whole Midwest. Someone called from Wichita and said the samples had been found. They were in Kansas. Sitting out on a loading dock.

"Are they being heated?" I asked. There was an edge to my voice that I hadn't heard for a while. "We don't heat our shipments," the voice on the other end replied.

After a pause, the voice asked, "What are in the boxes?"

"Water," I said.

"You mean ice, by now, don't you?" And he gave a nervous little laugh. "They should be there in a week. As soon as this weather breaks."

I slammed the phone down.

Two weeks later I found them on the loading dock at Boyd Hall. I knelt down with a crowbar and popped the metal bands. I pried open the lid. I was expecting to find broken bottles and dirty ice extruded among the Styrofoam chips. The legendary expansion of water when it freezes. Little explosions, one by one, in the backs of eighteen-wheelers moving over the snow lanes of the Midwest. Months of work gone.

But it didn't happen. The bottles were intact. Every one of them. "Alfred Russel Wallace," I said. "It must be."

We worked all night collecting and processing and preserving samples. Then we packed. When the helicopter arrived, Mike and I were sitting there in the midst of our gear, waiting. Tim and Dr. Yu were sleeping in the tent. They would remain at Vanda until after Christmas to collect more samples. I had my arms wrapped around a small wooden box. In it were the samples. These were going back to Canfield.

The flight plan seemed unusual to me. We would head west up valley and over the Asgards to Lake Joyce, pick up some barrels of waste from one of the camps, sling-load them down Taylor Valley, and then cut back to McMurdo. It seemed too indirect, too long. I just wanted to get to the mess hall for breakfast and then to the Hotel California for a hot shower and a good sleep. I wanted to mail the samples back to New Haven, where Canfield was waiting for them. As I climbed into the chopper, I was thinking of a soft pillow, a wool blanket, and a dark room where I would not hear the sound of the wind. I was thinking of how long it would be before the first data came in.

We landed at Joyce only long enough to hook on the load. The copilot had the controls. He tried to lift the 1,200-pound cargo straight up off the valley floor. It wouldn't budge. The helo just hung there, straining itself in midair. The pilot took the stick. He torqued the engine from ninety to a hundred percent. The helo inched forward, but the load dragged. The cargo master was saying, "You're dragging, sir, you're dragging. The load is on the ground, sir." The pilot was looking down at the shore of Joyce.

We landed again, jumped out of the helo, unloaded some barrels, put some lighter boxes into the sling, cleaned up the little mess we had made around camp. We took off again. This time we barely cleared the ground. I could hear the copilot over the radio, saying, "This damn load's still dragging us down." Finally we were steady at about a thousand feet, coming up the Taylor Valley. We could feel the sling like a pendulum beneath us. Every few seconds, at the end of one its swings, it would jerk to one side. You knew it was down there, moving, observing its own Galilean laws. No one spoke. Mike's chin was lowered into his parka. He was shaking his head slowly from side to side. "Why in hell are we doing this?"

You never know what to expect in these valleys. One minute you are looking out over ice fields and cirques and nunataks, all set against a perfect sky. The winds are at your back and the helo is holding level, racing along. You round the snout of a glacier with a carefree grace that catches in your groin; the helo tips a little to the left, then rights itself. You head off down a narrow chute of stone. You are high on the cliff walls, close enough to them that, if you could just open the window, you could breathe on stone a thousand feet up. Even though you think for an instant, "What would happen if this blade touched the rock up here?" you know it is hardly possible. Like the aerialist on his wire, hundreds of feet above the crowd, you know the pilot and the machine are in absolute control. You can hear that in the voices coming through the headphones. The voices are sure and confident. These are the best pilots in the world. Toward the east you are looking over a hundred miles of frozen sea and cloudless sky. The volcano smokes in the distance; its plume is a planetary flag.

Then it changes. A crosswind funnels out of the pass. It is a river of air raging down off the mountain. "Ah, power," Dr. Yu would say. But it is invisible. All you can see ahead of you is the utter clarity of the day. Without warning, the Huey is in the midst of it. It is a raft in a swollen stream, and it begins to pitch and roll. The pilot is gripping the stick, trying to keep it on course. Suddenly you remember where you are.

We were moving into heavy weather, into a mass of air that was coming off the Sound. The moisture was condensing, the way vapor from Acton Lake condenses on a cold morning. The pilot turned the machine toward a patch of blue sky that he saw off to his left. It looked as though it would take him out like a tunnel over the clear sea. But before he reached it, it closed in on itself. It was gone. We were flying in indiscriminate whiteness. We could not see the ground, we could not see the horizon. The ice and the sandstone that had been everywhere visible to us were now gone. The Ross Sea had disappeared. I felt cold sweat trickling down my side. I took my wool cap off, opened the collar button of my shirt.

Mike had his face pressed against the window. He was rubbing his mittens against it, instinctively trying to clear something away. He was moving his hands rapidly, in widening circles, the way you do on a frosty morning. From where I sat, I could see nothing. No horizon. No distance. No shadows. Over the radio, I heard the copilot say,

"Vertigo, vertigo. I think we're spinning." I was clenching my teeth, muttering "Jesus" into the tiny microphone on my crash helmet. It felt as though we were turning. The oscillations of the sling were becoming wilder. Its amplitude was increasing. It was throwing itself out from the body of the ship and pulling the ship with it. It seemed only a matter of time, a few seconds maybe, before something would break us, before the load would pull us down. I felt as though we were flying in a maze, in a labyrinth whose walls were only inches from the ship's blade. How will it happen? How will it feel? How will they find us? I was thinking these things, one on top of the other. I was thinking of the samples. They had taken a year to collect. Who would ever bother to look for them? They were only dirt.

The day you arrive at McMurdo, they tell you about zero visibility. If you are not listening, if you are nodding off exhausted after the long flight in from Christchurch, Erick Chiang or Dave Bresnahan will stop and reprimand you. "Listen up," they say, "this could save your life." The outside doors on every building at McMurdo are easily spotted. The huge freezer handles are on the outside. The doors are always unlocked. You can lift the handle and fall against the door out of the storm. Still, there are stories of people who have been lost on the main street of town, in the quarter-mile between the dining hall and Hotel California—people who have not been able to see their hand when the whiteout came. I heard of a man who was killed by flying plywood as he walked only inches from the Biolab door. When they found him, his arm was outstretched in the snow as though he had seen it.

You worry about a whiteout in the same peripheral way that you worry about crevasses in the ice fields, or in the same way that you worry about the Herc ditching at sea. You worry about it, but you think you will never see it. When they give you the dogtags in Christchurch and you throw them around your neck, or when you look up at Curtsinger's dark ocean on the wall, you think of those things for a moment. Little dangers. But then you forget. You lose yourself in the science. You lose yourself in the place.

I don't know how long we were flying blind. Or how long I was muttering to myself. Or how long Mike was clearing the glass. It could have been hours. I don't how long it was before the pilot caught himself, before he remembered something he had learned at Port Hueneme, something he had learned flying in the High Sierra. I don't know how long it was before he remembered to shout, "Up, up, take her straight up. We've got to climb."

I felt the Huey lurch. I felt the load shift beneath us. We had strained to get off the ice at Joyce. We had strained to get over the glacier into Taylor Valley. We were straining now to gain altitude. The twin engines were torqued to screaming. You could hear the *whomp, whomp, whomp* of the blade only a few feet overhead. But you could see nothing. Cream. Milk wisps and streamers rushing by the window. Beyond that, beyond the turbulent lamina that clung like hoarfrost to the skin of the ship and then spun off, there was nothing but a featureless opacity. I was becoming dizzy just looking at it. I was feeling hot and sick.

The ship was climbing, but it sounded as though it were coming apart. The rotors were whining the way they whine at shutdown. I thought they might stop. I was looking for signs, anything. A mountain peak, a shadow, my own reflection in the window. But there was nothing. There was light, and it was scattering all around, yet I was blind. The pilots were blind. I wanted to stop and huddle against something. I wanted to be at Bull Pass, in the mountains, with the snow coming on out of the west, with the huge boulder at my side. In the mountains I felt free. Here I could hardly breathe.

It happened all at once. It was like one of the transitions that water makes. One minute there's nothing but cold drizzle, the thin rain is an affliction; the next minute the roads lock and turn to slick glass. We had to be at ten thousand feet, but I was still imagining what the walls of a glacier could do to us. Then the whiteout disappeared as suddenly as it had come on. There was not a streamer or wisp of cloud anywhere. We were above it all. There was nothing but blue sky over the entire sound, over the Ross Sea. O'Keeffe's sky, impossibly blue above the steeple of the desert church. The slopes of Erebus were on fire with reflected light. The lands of Ross looked perfectly still. A quiet splendor. It was as though nothing at all had happened. Nothing.

I heard the pilot say, "We're at fifteen thousand feet. We're above the ceiling. God, it looks good up here." Then he turned to the co-pilot and said, "Radio Mac. Tell them we're coming in, tell them we're coming home."

We began drifting down through vaults of azure and indigo, through the scattering of small waves, blue skies, toward McMurdo. There was no wind. You could hardly feel the sling load beneath the ship's belly. In the distance, at the base of the volcano, the huge fuel tanks were burning silver. There was a line of turquoise water that ran clear, like an ancient river, far off into the Ross Sea.

For just a second I felt as though I were on a cusp, on a knife edge. At a critical point in one of those phase diagrams. But here, instead of water and ice, were terror and beauty lying side by side.

We tend to think of rainfall and snow as pure water. But they are really not pure at all. At the center of each droplet of water, each six-sided plate of snow, is a particle, a microscopic bit of matter, about which the fugacious water molecules can congregate and cling. For water does not coalesce easily into droplets. Just as a droplet begins to form, just as the molecules curl over on themselves to make the first shapings of a sphere, there is a compensating tendency for the molecules at the surface to fly back into space. Vibrations, rotations, and translations at the molecular level literally tear the nascent droplet apart from the surface inward. Pure water is without a center; it cannot sustain itself. It forms and vaporizes in an instant.

Fortunately there is no shortage of particles; wherever there is water vapor, there is rain or snow. The particles come from everywhere. From the continents there rises an invisible storm of mica, quartz, and bits of clay, calcite, gypsum, and hemitite dusts. Iron-rich particles from the Gobi Desert waft over Asia, down the crowded streets of Beijing, and westward through the vast Pacific. And there is the ash of forest fires, the voluminous debris of volcanoes—think of Mount Pinatubo, whose particles turned the skies of the Midwest lavender and autumn gold for years—the terpenes from every Blue Ridge spruce, the countless compounds of sulfur or ammonia rising from marshes and sea grasses, from the buoyed bodies of ocean plankton, from all the watery and landborne exhalations of the world.

It is at the surface of these particles, so vast in number and kind, so variegated in size and shape and hue, that the molecules of water gather and adhere. With something solid on which to cling, layers of molecules condense, build up one upon another, until a droplet has formed. Without particles, without centers, raindrops and clouds could not exist.

But it is not only these windblown bits of stone that bring the Antarctic world to a standstill. On the continent the temperature is low enough at ground level that clouds are composed not of water droplets but of tiny ice crystals. The crystals are prisms and mirrors, turning as they hang suspended, tumbling as they fall through the air. They bend and reflect the rays, send them back on themselves, turn them left and right, dally with them like signal mirrors. The result is that illumination is equal in all directions. There is no up or down, no sky, no horizon. Only the molecules and the soft strange light, and the sense that you are spinning, tipping over; that you are about to die. At one point in the helo, I wondered, "What nuclei did this?" Then I realized we were near open sea. It might have been sea salt. Just sodium chloride.

The pilot brought us straight in over the seal huts. He hovered thirty feet above the pad while a flatbed truck rolled beneath us. The cargo master gave the signal, and in a moment the load was in. The truck crawled away up the cinder road toward the Chalet. The pilot shifted the Huey a little to the right and the skids touched. In a few seconds the engines torqued down, the blades stopped. I unbuckled myself, stretched my arms wide, reached around to touch the samples to make sure they were still there. They were. Then I swung my legs over the seat and jumped out. The pilot was standing with his flight helmet under one arm. He was looking at the valleys; it seemed that he was still over there.

I walked up to him and shook his hand. "Thanks for what you did up there," I said. "I thought we wouldn't make it." I really didn't know what else to say.

When I turned to the helicopter again, I thought I was seeing things. Just inside the cab, bending over the box that held the trap samples from Vanda, dressed in wool pants and tennis shoes, as though he were a tourist, was Varner. I couldn't believe it. I had to look twice.

THIRTEEN

The Mailroom

I don't know what happens here, but after a short time everything above the Antarctic Circle seems chimerical, an invention of one's isolation. News of wars and revolutions, of diplomatic shufflings and intrigues—in fact, news of the whole frenzied twitter of civilized life—has about it the quality of implausible rumor. You realize just how distorted time and distance have become, how, in mere calendar months, whole decades, whole centuries, have sped away. You realize how pointedly strange so much of what we do is.

Admiral Byrd wrote, "The few world news items which Dyer read to me from time to time seemed almost as meaningless and blurred as they might to a Martian. My world was insulated against the shocks running through distant economies. Advance Base was geared to different laws. On getting up in the morning it was enough for me to say to myself: Today is the day to change the barograph sheet, or, Today is the day to fill the stove tank. The night was settling down in earnest."

Like the barograph and the stove, our world revolved around Varner's flume and Walt's phosphorus analysis and Dr. Yu's bamboo. Those were the things that mattered here. The latest fashion, the latest coup in the great world to the north, were blips too small to mention.

▼ ▼ ▼

McMURDO HAD NEVER LOOKED so good. We were too late for breakfast, but the chefs heard from Mac Center that we were coming. They had heard about the whiteout. They kept everything on for us. The scrambled eggs and the Virginia ham were sizzling on the grill. There were grits and oatmeal, toast and marmalade, a bubbling cooler of orange juice, steaming pots of coffee, doughnuts and breakfast rolls. You could smell everything. At Vanda, I had hardly eaten. In the last thirty-six hours I hadn't eaten at all. There was a cloud of steam rising from one of the warming trays. I held my hands in it, rubbed them together until they dripped hot water.

When I got back to the Hotel, I took off my parka and bunny boots, my windpants and vest, my checkered red hunting shirt. I tossed them into a pile at the foot of the bed. I walked into the hall-way, felt the soft carpet against my bare feet. Little gusts of warm air blew up from the vents. They made me shiver with pleasure. In the empty lounge, I jumped and touched the ceiling. I paged through magazines that were lying on the sofa. I wasn't reading them. I was looking at their colors. I was losing myself in their opulence.

I had lost weight. I could see it in my face. "Who is that skinny guy in the bathroom mirror?" I asked myself. My waist was thinner, my shoulders had broadened. I felt as though I were being shaped by the lakes and the mountains. I felt as though I were coming to some kind of flesh-and-bone equilibrium with the land. "There are men of a province who are that province. There are men of a valley who are that valley," Wallace Stevens wrote. It felt like that.

Perhaps I stayed in the shower longer than I should have. I washed my hair twice. I turned slowly in the hot stream that rolled over my back. The soap became vapor, became a dozen perfumes mingling with the ris-ing steam, with the sounds of water against the marble floor. Through the window, whose ice had begun to melt in the late-morning sun, I could see fresh snow on the upper slopes of Observation Hill, up by Scott's cross. On the pad, there was a helicopter hopping nervously from skid to skid, about to lift off. I looked away from it and raised my head again into the stream. I could feel the clays and salts of the valley being washed away.

I went back to the room. I pulled down the blind and closed the curtains. I ran my hand over the map, smoothed it a little. It looked as though no one had touched it since I left. I turned back the sheets and the woolen blanket. The room was dark except for an edge of light that fell beneath the window. The darkness was more than mere absence. It was substance. It was tangible. I could lie against it. I could let it cover me. I could feel it crest and settle over me like a great wave.

And there was no wind. No sound. Nothing moving. Nothing about to fall. I knew I would sleep. I knew the dreams would seem real.

When they took pictures of us, we were always together. The three of us. In the stream in the dark woods, you can see Eugene bending down in the current. There is a silvery braid of water in his open hand.

You can see Elizabeth's face. It is thin. Her legs are thin and she is trembling there in the shade, the water crested around her knees. At the Jersey shore, we are standing in breakers. We are holding one another up. There is a wave coming. It is a huge mirror, curling, concave. It is hurling the light. You can see the light in the strands of Elizabeth's hair. In the right-hand corner of the photograph there is a shadow shaped like the wide brim of a woman's hat. Off to the left you can see a white shirt, like a streamer fluttering. Much later, at the Cape, we are seated on Adirondack chairs. Elizabeth is in the middle. Before us everything is blooming in tangles of green and pink, shells scattered near the stone wall. Eugene is thin, muscular, tan. There is a crimson shirt across his knees. My mother called these the "waterpictures."

On Beacon Hill, we were gathered around the bed. There was Gregorian chant, gold crucifixes in the room. You could see the weathered copper of the roofs of Boston; it was oxidized green. And in the room Eugene motioned to us as though he were sweeping us up with a grand gesture of his hands. He could not speak. He could not move, except his arms, his fingers. We held him between us, one thin arm over my shoulder, one over Elizabeth's. My mother seemed confused at first; then she understood. She took the camera. There was a glistening skin of water over her eyes. Eugene was smiling. In the photograph he looks almost transparent, aqueous.

He motioned to the bedstand, to the book there. It was Henry Adams's *Mont-Saint-Michel and Chartres.* I slid it toward me, opened it, and read to him. I began somewhere in the middle, it didn't much matter where. He just wanted to hear the words. He had always loved this book and anything else Adams had written. His eyes were closed. "'Who has ever seen!'" I read. "'Who has ever heard tell, in times past, that powerful princes of the world, that men brought up in honor and in wealth, that nobles, men and women, have bent their proud and haughty necks to the harness of carts and that like beasts of burden they have dragged to the abode of Christ these wagons, loaded with wines, grains, oil, stone, wood, and all that is necessary for the wants of life, or for the construction of the church.'"

When I looked up he was asleep, snoring gently. It was afternoon and there was snow on the steep roofs of Boston. In the morning he was dead.

All of this happened. In the dream at McMurdo, it happened again. There was no wind. But in the dream the "waterpictures" trembled in my hands.

When I awoke, I was still thinking of Eugene, how one night over dinner, in a small bar at the foot of Charles Street, he showed me a paper he had just written. It was on liturgy. On what it meant to be elevated in the presence of God, to be "edified," as he put it. He reminded me that edification comes from the Latin *aedificatio*, to build up, to instruct. Edification, he had said, was a state of mind, tied in with a sense of the sacred, with feelings of awe, with that "nudge toward the transcendent." An edifying liturgy occasioned tranquillity. It occasioned calm and centeredness, a sense of connection with something larger than the self.

I told him how I had once felt that. Had felt it in church with the choir and the incense and the long shafts of light. But no longer. And he said, "But you know exactly what I mean." He had his hands folded before his face. His eyes were bright. He looked strong and healthy, filled with plans. "You feel it down there, on the continent. Every time you go. It's the same mystery, for both of us. It's all around."

The next morning Varner brought coffee up to the room. He sat down and told me everything that had happened. When he'd gotten back from Lake Miers, he'd been physically exhausted. He'd told Walt he wanted to spend the next day alone. He didn't want to see the lab; he didn't want to see anyone. He sat up here in the dark with the drapes pulled, and smoked packs of cigarettes and drank beer. He began to think about where he was, about Carolyn, about the kids, about his sick father. In the absence of mail or any kind of communication, he began to imagine things. There would soon be ice on the roads back home. Maybe his son or daughter would have an accident this year. Maybe his wife would leave him. He wasn't certain about this. Maybe his ailing father would die. Even though Varner's father had told him, "I want you to go there, and if anything happens, don't even think of coming back," Varner was uncertain about what he would do.

Once he had seen the valleys, he said, he was not sure he wanted to go out there again. "Just rocks," he said. "The way people used to think about the Alps." He told me he thought I had fed him and everyone else a line when I said how beautiful the valleys were. For him they were just *there*. Indifferent at best. They had no special value. He thought I was reading something into them. "You're inventing the damn place," he said. "It's all in your mind. It's ice and stone. It will be there forever."

He seemed to have a metaphysical distaste for it, the kind of feeling that some people have for the sea: threatening, all subconscious. There was nothing he could get hold of, nothing he could find there to draw into himself. It all repelled him, or at best it was immiscible, would not join with anything that he considered his. He hadn't expected this.

Varner had spent most of his time at McMurdo alone. Once he'd gone with the chaplain over to Scott's hut, browsed among the faded biscuit crates and the seal carcasses still there on the floor grinning up at you. But that was all. When he'd heard about the flight, he'd put his name on the manifest, ready to head off for Christchurch and Akron. He was not happy about himself, but he was leaving nonetheless. Somehow the work would get done without him.

As it turned out, the same plane he was scheduled to fly out on had brought in a load of mail. The people at the Chalet had sorted the mail and brought it up to the Hotel just as he was flinging the last shirts and socks into his flight bag. He checked his pigeonhole on the second floor and drew out a letter. He didn't have time to read it. He knew the "Delta" was loading up outside the Chalet. He raced downstairs and nearly collapsed against the wall of the huge orange cab as it began to roll up the hill out of McMurdo toward Scott Base.

When he arrived at the waiting room at Williams Field, he reached into his parka and took out the envelope. Its return address read. "Carolyn Varner, Mayfield Ave., Akron, Ohio."

It was a long letter and he read it slowly. When he got to the last few lines, he traced them with his finger. "You know," Carolyn wrote, "at the Cleveland Airport I thought you weren't going to get on the plane. But I'm so glad you did. I really think you needed this. I know everything will work out. I'm proud of you. I love you."

He read the letter again. He was sitting in one of those hard-backed chairs they have at Williams Field and he was looking out on the runway. There was a fine snow kicking up from the ice, like diamonds being scattered around.

The captain began his briefing. It was the usual stuff: the life rafts, the inflatable cold suits—"lobster suits" they called them—how the Herc could fly on only one of its four engines, how it could ditch at sea. Varner had folded the letter away into his parka. He was facing the captain, but he wasn't hearing a thing.

The whole journey was playing itself out again in his mind: the medical tests, the shots, the skeptics who had told him they would never have expected this, not Antarctica, not from him; and the students from the newspaper who had come to interview him, who had told him how proud they were that one of their teachers was doing this; and his father, seated there in the big family living room in Youngstown, the look he had given him when he said, "Don't you even think of coming back for me." These and the flight from Cleveland and all the doubts he had had on the way down. The hydrologist who had told him there was no way he would ever measure stream flows with a flume. The whole thing would ice up. The numbers would be meaningless.

He thought of these things and of Lake Miers and of the stark reality of the place—something he had never imagined—and of his long seclusion at McMurdo. How strange Walt must have thought all of this was: his decision to leave, and now this—this letter. What should he do?

People were moving toward the door. They were moving out single-file with their flight bags. Down the three wooden stairs that put you onto the hard-packed snow near the runway. They were stopping for a second to look up at Erebus. It was an unconscious pause. Something people just did naturally.

When he got to the bottom of the stairs, he waited. He looked at the volcano. It was a perfect day. Even here, he had never seen such blue sky. When the captain walked by, he cleared his throat and spoke to him. "I think I'll fly out later," he said. "I have some work I think I should do." The captain nodded and Varner boarded the Delta, which was about to return to the Chalet.

By the time he had finished his story, the saucer was littered with cigarette butts. He was beginning to tamp out his Marlboros in the coffee cup. There were cold ashes lying everywhere on the bed. I said simply, "I'm really glad you're here." It was a big understatement. I was elated. Both for myself and for him. I was happy for all of us.

▼ ▼ ▼

Later that day, Varner said he wanted to show me something. It was up at the Berg Field Center. We put on our parkas, went outside, walked a few hundred yards up the steep cinder road. The first floor of the BFC contains wire cages. Ten feet wide and fifteen feet deep, they run from floor to ceiling on either side of the wooden isle. On each door is a combination lock. We had been assigned a cage toward the rear of the building. We stored our food and our equipment and our camping supplies there.

The first floor is usually dimly lit. Only near the door, where they keep the Nansen sleds with their polished runners, is there much light. We walked toward the back, into the semidarkness. I was running my fingers along the screen to steady myself. Varner seemed more confident of his footing. We stopped and he began to twirl the face of the lock. It clicked open and he pushed the door inward. We entered the cage. He picked up a flashlight that was lying on top of a crate and flicked it on. He swept it through the darkness, until it fell on something wooden. At first I was confused. I wasn't sure what I was looking at. "Let's take it outside," he said. "It's not very heavy. It's only plywood."

We put it down in the cinders outside the door. A large plywood box, open at the top, it was four feet tall, about six feet wide. In front, it was shaped like a V. Varner was talking. He was saying, "The idea is to position it so that the V-notch faces downstream. Then, as the water channels itself through the notch, you measure its depth. If you know the geometry and the water depth, with a calibration curve it's possible to measure how much water the stream is discharging, how many gallons or liters per second are spilling through the notch. The problem is getting the whole stream into this box. I don't know whether we can do it, but it's worth a try."

"It's beautiful," I said. "You don't how beautiful. I think it will work."

He was looking at it with obvious pride, as though it were a finely crafted sled, as though it were one of the polished Nansens inside the door. He had his arms folded over his chest. He was moving cinders around near the base of the flume with the toe of his bunny boot. Then he glanced up at the mountains in the distance, at the Society Range, filled with light. "I can't wait to see what it does over there," he said. He pointed across McMurdo Sound with his glove. For a second I hardly recognized him.

On our way down the hill I was singing to myself. Maybe it was the way the light had turned the mountains silver. I don't know. But I was thinking again of the poem Borges had written. The poem of the dropped coin. I had read it so often I remembered every word, and the words, in a way, were about the metals, the metals in the lakes. I repeated them to myself:

Cold and storm-threatening the night I sailed from
Montevideo.
Coming round the Cerro,
I flung a coin from the upper deck
And watched it flash, then sink into the murk below—
A thing of light swallowed up by time and darkness.
And through me went a sensation of having committed
an irrevocable act, of adding to the history of the planet
two incessant, parallel and perhaps infinite series:
my own destiny compounded of anxieties and love and
pointless struggles, and the destiny of that metal disk
which would be borne by tides into the soft chasm
or out to remote seas still silently gnawing
at Saxon or Viking spoils.
Each moment of my sleep or waking
is matched by another of the blind coin's.
At times I have felt remorse,
at times, envy,
of you, like us, walled in by time and its labyrinth
without knowing it.

Sometimes I felt there was no better expression of what we were doing here. That each of the metals—the manganese, iron, cobalt, nickel, copper, zinc, cadmium, and lead—weathered out of the valley's bones by the trilling streams, was like the coin, flashing for a moment where the water touched it, then "swallowed up by time and darkness" in the chasm of the lake. But by how much time? For each element it was different. Each had its own destiny, each traced its own curve through the world. For each there was a history, a fate going on outside our fates.

"Walled in by time," Borges had said. By how much time? Varner's flume would let us measure this, would let us calculate just how long things remained. These fates going on outside our fates, these journeys that paralleled our own, were coming into view.

"How can the world be like this?" I wondered, as we walked

toward the Biolab. "How can the world be like this that we should know it, that we should have this sense of its workings?"

▼ ▼ ▼

The foyer smelled of hot chocolate. It smelled of herbal teas. When the door opened it smelled of diesel fuel and cold air. As I walked farther into the building, there was a taste of mixed reagents on my tongue, then the sweet scent of isobutanol. Walt had been working on the phosphorus last night.

His notebook was on the desk. There were neat columns of numbers up and down the pages. Walt's notebook was all business. It was different from Canfield's, where you didn't know what to expect, where a bunch of numbers would trail off into a knot of differential equations, followed by an exclamation—"Beautiful, windy day by the Onyx River. I feel great!"—that would just jump right out at you.

With Walt you knew exactly where things were. You knew when everything had happened. I paged through the whole phosphorus story. It was laid out in clearly written numbers. There were only a few words, but it was possible to see where he was struggling, where the data were making no sense at all. At times he was getting phosphorus readings that would have made a detergent maker blush. They were off scale and he knew it. The pages of the notebook were filled with frustration. You could see that in the hundreds of data points. You didn't need an interpreter.

Then last night he got it. He started the whole analysis from scratch. He had ordered all new reagents from Christchurch. He had put a rush on them. Which meant weeks. Yesterday they came. Bottles of acids and alcohols, the molybdate salt and the potassium antimonyl tartrate. All of them were the highest grade of purity. At first he had done only three standards and two samples, the way you do when you are not quite sure of yourself. But these had been perfect. Just a hint of molybdenum blue swirling along the silver walls of the flasks. The three standards fell on a perfect straight line. Beer's law! And the samples lay between the highest and lowest standards, as though he had planned it, as though he knew what to expect of Lake Miers. Maybe he did. He had been there, had touched the lake; he had felt its waters flow through his fingers.

He ran the whole profile, top to bottom. There was hardly any phosphorus in the lake at all. Ten micrograms per liter, more or less, except near the bottom where the water and the sediments touched. The analysis looked good. The numbers were well above the detection limit, but they were finally low enough to be credible. I closed the notebook and looked up. Out in the foyer there was a big commotion. There were reports that two thousand pounds of mail had arrived. I ran back to the Hotel and waited.

My room was just across the hall from where they sorted the mail. I had finished packing up the particle and water samples for Canfield. With any luck we would have some preliminary results before we left. They would go out on the evening plane to Christchurch. I had a letter for Wanda that I had been working on every night at Vanda, and cards for Dana and Kate. On the envelopes for the girls, I stamped a large emperor penguin. On the broad white belly under the chin I printed DANA on one and KATE on the other. Below my name 1 wrote the return address as "Antarctica." I knew they could find me.

In the hallway I heard the sound of the mail. It was unmistakable. The long *ssshhhhh*, the slam of doors, another *ssshhhhhhh*, the mailbags being dragged along the floor, joined with the tuneful voices of the people from the Chalet. From my room I could hear them sorting it, speaking the names of their friends as they put the envelopes into boxes labeled A to C, D to F, G to I, and so forth. I could hear someone announce, almost disappointedly, "Here's another one for Anna. I can't believe it. That's five. And I haven't gotten one yet."

Sometimes there would be a burst of excitement when a large package was pulled from the sack. "Wow, Bob struck it rich on this one. Look at the size of this thing. I wonder what it is." I could imagine someone in a bright plaid shirt holding a cardboard box up to his ear, shaking it, listening to things thumping and scattering around inside. "Hmmmm, cookies. I think I can smell them. And who knows what else? Jesus, it's civilization!"

I heard someone say "Varner," and then "Varner" again. Later I heard my own name spoken. "Isn't he at Vanda?" someone asked. "Shouldn't we put his mail on the chopper and send it out?" "No. I think he's here," another voice replied. "Let's put it in the box."

People were coming up the stairs of the Hotel California. There was a crowd gathering in the hall. There was laughter and joking. Friends who hadn't seen one another for a day or two were exchanging greetings as though it had been months. I heard the whispers of parkas and windpants moving just outside.

I waited until things had calmed down, until the last footsteps had trailed away. Then I opened my door and went over. Whatever detachment I had been feeling for the world north of Hut Point evaporated with the first step into the mailroom. Suddenly I wanted letters. I wanted to hear about home-cooked meals and report cards. I wanted to hear about Halloween and the first snowfall, about the morning coffee runs to Acton Lake, about the new doughnut shop. I wanted to imagine the living room and the study, the clutter of toys, the backyard as it stood under the night sky. I wanted to smell the scent of the house as it clung to the written page.

There is always a feeling of trepidation as you approach the west wall of the room, where the boxes are neatly stacked. You peer inside, take out a handful of letters, and begin to sort them, looking for a familiar script, for a type of paper, looking for your name. If there is nothing, your stomach sinks. Then you think, "Maybe they put them in the wrong box." So you sort through everything, through all the mail, through every civilian letter that has come to McMurdo. Then, if there is still nothing, you return to your own box. Maybe there is something in there that you didn't see. Maybe you just missed it. No matter that the thing is smaller than a shoebox. No matter that you can see right into it at eye level. You dust it out anyway. Hoping. Desperate.

It turned out to be a good day for letters. I took everything to the room, lay on the bed, opened the envelopes, and began to read.

Wanda's letter was what I had been hoping for. I could smell the kitchen as I read it. I could see the sun dropping off beyond the slumbering fields. It felt as though she were right here, as though Dana and Kate were playing upstairs. She wrote about Acton Lake, how she had seen it steaming one cold morning, how the mists seemed to rise from it in tall columns like stalagmites that broke and dissipated in midair. "It's a treasure," she said. "Every day it's different. Today there's ice all the way to the dam, all the way over to the lodge." She told me she had seen seagulls that had flown all the way down from Lake Erie. They had gathered near the far shore around open water.

She sent along a poem she had written. It was about water and time and work.

We try to trap
The essence,
The molecular dance,
In rates and flows.

Diagram the icy stillness.
But all the while
Air rushes
Knife cold, indifferent
Through our hands.

I opened the envelope from my mother. It was carefully written in her meticulous hand. The letters drifted in delicate curvatures and loops and volutes across the page. Each line was perfectly horizontal. The grammar correct. The way it had always been, back as far as I could remember. There was news of the day—of the snowstorm that had come a few weeks ago, that had caught her off guard when she was at Sunday mass. It must have snowed and snowed and snowed, the way I remembered it once. "Like the Great Snow," she said, a whole foot in an hour. "You should have seen the Cathedral, it hasn't been that beautiful in years."

She wrote of sports, of the Pirates and their prospects for next year now that the pitching would be better; and of the Steelers, and how they weren't doing quite as well as she had expected this season. "I miss Bradshaw," she said. She wrote of how the mills had fallen on hard times, how most of them were gone. "I remember the soot falling from the skies," she said. "The laundry in the backyard used to turn black. Your father would get so mad when he had no shirts. But that's all changed now." She wanted to know if I had been keeping track of the Antarctic ozone hole, if I had been following the news in the paper. "Do you need to wear special glasses in the field this year?" she asked. "I hope you are being careful with your eyes."

Then she told me she had found an old book in the attic. It was by Richard Halliburton, an adventure writer she had always admired. She'd read me his stories when I was five or six. It must have been in the living room, in the evenings, before we had our first television set, because when I think of Halliburton I think of the way the lace curtains hung, how they opened out onto the dusk, to the winter sun sinking in shades of red behind the hill. I think of her with the book spread across her lap, the way it fell open, the way the pages lay flat against her dress, with the Infant of Prague behind her and the standing lamps casting their thin, crepuscular shadows on the carpet.

Things then were just beginning to condense for me out of some nebulous cloud. When she read to me, there was more than just the sound of her voice, her slightly formal presence—the stiff white collar against the navy blue dress—across the room. There were the words themselves, the images that the words and her voice created. I was beginning to imagine the way the world might look beyond the confines of the house, beyond where the cobbled streets and the trolley tracks ended, beyond where the hills touched the sky. She would call me over to look at the pictures: Halliburton, confidently standing with his ice axe on the Mer de Glace. Or Halliburton near the Spencer Glacier, between Seward and Anchorage. Or Halliburton with his sled dogs or his elephants, or with the gleaming marble of the Taj behind him. She had been intrigued by his disappearance during a storm in the South China Sea. No trace of him or the Chinese junk with its "iron bamboo" had ever been found.

She spoke about Eugene, about how terribly she missed him. How it didn't seem right that he was gone. Her own son gone before her. It challenged her faith. Toward the end of her letter, almost as an afterthought, she told me she had slipped and fallen in the snow on her way back from church. "My leg is a little swollen," she said. "But your old mother will be all right." She wrote this in such a matter-of-fact way that I didn't think about it again.

"Please bring me a shell from your valley," she finished.

FOURTEEN

The Gallery

The young student was speaking with such intensity. "How can you compare art and science? I'm an artist," he said. "I have friends who are artists. I have friends who are poets. We are creative. We imagine things. We feel. You scientists, you are collectors. You gather things. There is no feeling in what you do. No passion. You are analysts. You are rationalists." He was throwing these words out as though they were epithets.

I left the classroom depressed. It had been my first seminar as a new instructor. I was lecturing on the similarities between science and art. No one really believed what I was saying. There were some polite smiles, some nods of the head, but little more. No one in the class had even heard of Bertrand Russell. When I read Russell's quote about art and science being the crowning glories of the human race, most of the students seemed to agree with only half of that. "Science. A crowning glory?" They hadn't thought of it that way. They hadn't thought of science and art joined together in the same sentence, much less united by a common sentiment. Why should they?

Everything they had ever been taught suggested that art and science were radically different, even antithetical. By temperament, you were either an artist or a scientist; you couldn't be both. You couldn't even relate to both. Somewhere, early on, you knew which one you were. You quickly banished the other from your mind, paying it only enough grudging attention to pass a required course or two. Science and art. They were separated by a deep valley. A river of ice had cut between them long ago, longer than anyone could remember. They would never be joined. You found yourself on one rock outcrop or another, surveying the chasm between.

▼ ▼ ▼

THE OFFICERS' DINING ROOM had been converted for the evening into an art gallery. It was, so to speak, the First Annual McMurdo Art Exhibition. Maybe it was the first art exhibition in the history of the whole continent. Everyone at the station—scientists, painters, plumbers, and bureaucrats—had been invited to submit work, and many had happily done so. Everything was welcome. Nothing was refused.

The voices inside were hushed. You could hear violins, the sound of wine being poured into plastic glasses along the linen-covered table. People spoke in whispers. They stood before the likeness of Mount Erebus, done in pastels and swirls of pink, as if before a shrine. There was a sense in the room that you were witness to something being born, to something coming into existence. That you were witness, as my student would have said, to creativity, to the passion and imagination of art. That you were in the presence of an intense personal vision.

I was running my hand over a metal sculpture of the continent. It was massive, done in poured lead, perhaps ten feet across. The surface of the metal was not perfectly smooth; it undulated in peaks and declivities, and the peaks caught the light and reflected it back as tiny brilliances, almost like farmhouses seen at night from the air. I closed my eyes and traced the whole shape, my palm moving slowly, turning a little against the lead, feeling its way through the far-flung emptiness of Queen Maud Land, then sweeping north and west into the Ross Sea. My thumb felt the fluted coastline of Victoria Land and I thought I touched mountains and valleys and, in the deepest recesses of the metal, lakes and rivers. I held there for a second, let the feel of the metal run up my fingers, then I moved my hand toward the peninsula, where it thins and fissions into island fragments, spinning away from the ice.

When I touched it, I wondered where the atoms beneath my palm had been. Whether one of them, in its long journey through the world, had been to Taylor Valley, had come rushing down the unnamed creek that hugs the base of the Canada Glacier and had been issued under the thick ice of Lake Hoare. Or whether one of them had once been an impurity in the great tin organ at Saint Petersburg, on that bitter day when the pipes turned to dust, when the temperature of the city sank below the transition point. I wondered how long these atoms would spend as sculpture—how geochemically rare that was—before resuming their more usual pursuits. I wondered whether I would meet any of these atoms again.

Across the room, on a small stand, was another piece of sculpture. It was called "Lee's Karma Indicator." It looked like a weathervane.

But it was constructed so that it could indicate only one of two states, GOOD or BAD. Nothing in between. Varner was over there chuckling to himself, turning the arrow from one position to the other as if he were expecting to feel something. Then he set it on GOOD and held it there for a long time, the hint of a confident smile on his face, as though he were happily imbibing a subtle fluid.

In one corner a crowd had gathered. They were staring up at a drawing hanging on the wall. There was animated conversation. People were pointing toward one part of the picture or another. On some faces there were smiles; on others there were pensive expressions. I walked into the crowd. I could just see the drawing from where I was standing. It was titled *After the Treaty*. It looked as though it had been done quickly, in pencil, maybe the night before. The sketch was large, about five feet by three feet, and was affixed to a piece of cardboard.

You had to look carefully. On either side of the Biolab, which was rendered as a tiny, one-story cube, was a slender, stately high-rise, glass-fronted, impeccably modern. In their presence, the Biolab seemed antique, a museum piece. Laughably out of date. Spanning the two towers was an arch that read NSF INC. HEADQUARTERS. The main road from the Chalet to Williams Field was a four-lane paved highway, two-tiered, with separate levels for incoming and outgoing traffic. Lined along it, bumper to bumper, were eighteen-wheelers, motorcycles, cars of all kinds. The highway rose up and curved in cloverleaves that swooped over to Scott Base and up to the cosmic ray facility on Observation Hill. There was a large road sign pointing up to Scott's cross. Like highways built into the desert, this one came to an abrupt end at the McMurdo International Airport. There were large hotels in the center of the town itself. The dusty streets had been replaced by fashionable malls and concrete squares. By Scott's Hut, an ornate sign was planted in the cinders. It read, COMING SOON, PIZZA HUT. Offshore, where the Coast Guard icebreaker *Polar Star* docks, there were huge piers against which a dozen splendid cruise ships were anchored. Little figures in formal wear filled the decks. I must have studied the picture for ten minutes. Every time I thought I had discovered all of its details, I would find something new—a little travel agency, a nuclear cooling tower nearly hidden behind the white foothills of Erebus, a 747 circling, waiting its turn to land.

The paintings in the gallery were mostly watercolors. They were done in pale shades. Soft pinks and blues graced the drab walls.

Erebus—which on this island has something of the same influence as Japan's Fuji, a constant presence, a thing of wonder—was sketched in dozens of shades in all its stages of illumination, at all times of day, and in all seasons. I thought of Monet's fields, his poplars and stacks of grain, the way the light transformed them in the French countryside. Because light could reinvent the world over and over, moment by moment.

There is no Gallery of Science. Not here, not anywhere. Somehow we have come to think of science, at its best, as discovery. We think of the scientist as someone like James Clark Ross, who, setting sail upon the waters of the unknown, comes, perhaps at great sacrifice, into the Land of Light; who nudges his ship through ice floes and howling winds into the safe harbor beneath the mountain; who gestures and points and names everything within view. This is discovery: coming nobly and with great fortitude and perseverance and with no little wit and energy upon that which already exists, upon that which, however hidden or far away, is already with us. Just as Ross came upon these islands and ice shelves and seas, so too did Dalton, we think, come upon his atoms, Rutherford upon his nucleus, Henry Frank upon his flickering clusters deep within the structure of water, and Robert Garrels upon his cycles of carbon and oxygen endlessly turning through geologic time. We think of the scientist as we think of Ross and Scott, as the discerners and revealers of what is actually out there in the capacious provinces of the world; of what is waiting passively to be discerned and revealed. We think of the scientist, at best, as discoverer. It is the artist alone for whom we reserve the word *creator*, for whom we offer the quiet observances and protocols, the diminished voices, of the gallery.

Perhaps this distinction is too simple, perhaps there is something about science we are missing.

When Henry Frank and Marjorie Evans published, in 1945, in the *Journal of Chemical Physics*, their paper, now famous among physical chemists, "Free Volume and Entropy in Condensed Systems," they introduced to the world the concept of "icebergs"—not as great barges and floes of ice, but as tiny arrangements of molecules, microscopic "icebergs." At first they used the word almost apologetically, as though they felt it was too metaphoric or imagistic to be a proper scientific term. But on page 520 of the *Journal* they dropped the quotation marks around the word with the following explanation:

> Throughout the rest of this paper we shall use the word iceberg, without quotation marks or apology, to represent a microscopic region, either of pure water or surrounding a solute molecule or ion, in which water molecules are tied together in some sort of quasi-solid structure. It is not implied that the structure is exactly ice-like, nor is it necessarily the same in every case where the word iceberg is used.

With this footnote, something important happened, something monumental even. Frank and Evans had discovered icebergs, had planted the flag of discovery on some ragged, wind-tossed coastline never before seen. They had come into the territory and given it a name.

Was this really a "discovery" in the sense that we speak of Ross's discovery of this island? After all, there was nothing tangible about these clusters, nothing you could get hold of or lay claim to. The clusters came and went, came and went—a hundred billion lifetimes in a second. There was nothing to see, nothing to point to, the way Ross had gestured heroically toward Erebus or toward the mountainous coast to the west. There was nothing to bump into, to steer around or sidle against, nothing to descry. So had Frank and Evans really "discovered" icebergs, or had they, in some deep and personal sense "created" them—created them in the very way that we reserve for art alone? Had they brought them, root and branch, into the light of being? Given them shape and form and title the way Brancusi shaped his *Bird in Space* or his *Mlle. Pogany*? Which?

I remember discussing this very question with Walt, my angry student. It was near the end of the introductory course he was taking with me. It was back when he was still questioning whether science could ever be anything more than a dust-dry collection of facts, more than a mere exercise in reason. We had been talking about Rutherford, about how he had scattered charged particles off sheets of spun gold; how he had measured the scattering angles with the help of Geiger; how he had been overwhelmed by what he had seen, by numbers on a page. There was only one conclusion to be drawn: that the gold atoms had a dense region of positive charge; that each had a nucleus.

After class, Walt came up to me and said, "But that was not a 'discovery.' That was an act of imagination. He made it up. It's as if he composed the nucleus. That was the word he used: "composed." As though the nucleus were a study in protons, a composition in dense matter, a rondo in bits of charge. I hadn't thought of it quite that way. I walked home thinking about what Walt had said, and how it seemed to make great sense. Rutherford's nucleus was a composition. The same was true of Arrhenius's ions. They were a composition, a mental portrait of how an aqueous solution of sodium chloride must be, at the most Lilliputian level, if it was ever to conduct an electric current. And icebergs. They too were a composition, a pure creation, a work of art.

There was water dripping from the trees as I walked home that evening in Ohio. A drop fell on my wrist. It warmed a little from the warmth of my skin. The hydrogen-bonded icebergs that formed and re-formed in the droplet began to disappear, became fewer in number. I could feel the ions of sodium and chloride quicken their pace as I walked, as the water broke away from them, from their charges, returned to the air.

Somewhere in his writings, the philosopher Karl Popper tells a story about Galileo. It is a story about the way we view the world, about the way we view scientific theories. Galileo argued that the Earth really moved about the sun. He did not say, "Here is a way of thinking about the motion of the Earth, a way that will make your calculations easier than the old ways." He did not say, "Here is a new model. I think you will find it useful for making calendars." No. Galileo said that the world was put together in such and such a way, and that the heliocentric theory of Copernicus, for which he was an impassioned apologist, was a portrait of reality, a composition in planets and space and movement. It was not merely an instrument of prediction. It was more than that. It was the truth.

According to Popper, Cardinal Bellarmine and the whole hierarchy of the Catholic Church would have been quite pleased had Galileo been willing to teach his system of the heavens as a mere mathematical model. After all, Jesuit missionaries had been using the Copernican theory in China for decades. Galileo was unwilling. The Earth really does move, he said; it moves around a central Sun, it moves with the other planets. It is not stationary. It is not the center.

The conflict here, between Galileo and the Church, turned on different views of what a scientific theory is. Galileo, the "realist," held that a theory was, in a sense, a sketch of the world, of some region or domain of it. Bellarmine, the "instrumentalist," claimed that a theory was only an instrument, a convenient device for predicting and explaining what we see around us. It is not a picture of what is actually there at all.

Popper said that most physicists, especially in the wake of the quantum theory, had become instrumentalists, had come to view physical theories as convenient fictions, or perhaps as ethereal machines designed to "crank out" correct answers. What irony! Contemporary science had ultimately allied itself with its ancient antagonist—the seventeenth-century Catholic Church of Bellarmine.

I spoke with Henry Frank about this once. I wanted to know what he thought about this long-standing, but now somewhat forgotten, dispute. Frank was not a quantum theorist, but he had grown up as a young student in the revolutionary times of Einstein, Bohr, and G. N. Lewis. As a physical chemist working on aqueous solutions, he had had ample occasion to apply the quantum theory to problems of water structure. I half expected him to proclaim himself an instrumentalist. But he surprised me.

"You know," he said, "there has always been enough philosophy in the problem of water to keep me busy. But if I had to bet, I would say that our theories, if they are good ones, pretty well describe what is out there in the world. Take the icebergs. You can't see them, but water sure acts like they're there."

It was refreshing to hear this. Henry Frank was a Galilean realist. He was saying, "If a theory isn't right, you know it. It just clashes with reality. But that must mean that there is a reality for it to clash with. I think the microscopic icebergs are real. I think our theories point to something real."

From the art gallery overlooking the frozen sea; from the shores of the ice-covered lakes; from the stone passes of the mountains and the narrow laboratory isles; from the nameless cinder streets of the town, the same question rings: What is this world? What is this world that we should know it? That in its frenzy, in its exuberance, we should discern, if not visually, then in imagination, if not with microscope in hand, then with equation and icon and conceit, its hidden structures and motions—the chase of its electrons, the deft swimming of its ions, the breathlike coming and going of its water clusters, its microscopic icebergs and bonds, forming like dense fogs, then clearing, then forming again, over and over, without cease. And that those structures and motions should in some way be real and true, for though we cannot see them or discover them as Ross discovered this sea, the world answers our calls as though they were surely there.

What is this world that we should know it? That out of the swirl and chaos of its sudden storm, out of the torrent of its silt-dark waters and the hues of its autumn leaves casting shadows on themselves, out of its summer corn exhaling mist into the dawn air, out of the rush of its objects, large and small—that out of all of these we should catch a glimpse here, a glimpse there, a pattern, a small epiphany of order, and that we should scribble it down, on napkins and shreds of paper, in the timeless fraying pages of journals, on anything that will make it last—this vision.

And who are we that we should know these things? That out of the mingling of water and stone, out of the touch of sunlight, out of the carbon drawn in long chains, out of the mats of heme, the iron and manganese, the calcium and magnesium of ancient seas, the seas of our life's blood; that out of the helices and rings of matter, we should dream these dreams—this mysterious deep time we cannot fathom, only measure; these cycles of matter we cannot control, which pass through us, which link us irretrievably to all that is. "Steep yourself in the sea of matter, bathe in its fiery waters," Teilhard de Chardin said, "for it is the source of your life."

We were scheduled to fly out to Lake Hoare the next day. There were reports in the gallery that the valleys were warming. The Kiwis were tracking the Onyx as it moved toward Vanda, thawing and freezing as it went. They had set up a lottery. The winner would be whoever came closest to guessing the exact time when the river reached the weir. So far the Onyx had not reached Bull Pass. But the appearance of liquid water in Wright Valley meant there would soon be water everywhere. We had to get Varner's flumes in. Spring was coming to the land.

When I went outside and glanced over at the Sound, I saw Mike coming up the road toward the building. He was carrying a long pole in his arm. When he got to the foot of the wooden stairs, he said, "I got it working. It's a backup. Just in case we have trouble with Varner's flumes."

It was the current meter we had brought along. It was shaped something like a tiny submarine, with little fins on the back. Mike had attached it to the pole. The idea was to dip it into a shallow stream and let it spin. As it spun, it clicked, and you listened to the clicks in a set of earphones that were connected by a wire. Once you had the number of clicks per minute, you could read the flow rate of the stream off a calibration curve. If you knew the stream's geometry, if you knew its cross section, then you could measure the discharge rate. It certainly wasn't as good as a flume, but it was very simple, and it was portable. This last feature, we thought, might be important.

Mike was looking at it as though pleased with himself. He had his arm stretched out and he was turning the pole very slightly, positioning it in some imaginary stream, squinting like a surveyor. "I can't wait to get over there," he said. "I think this will be our most difficult campaign. I had to pack up things for Hoare, Fryxell, and Vanda that I didn't know existed. I don't know if we'll be able to fit them all into one Huey. If we can finish this in the next two weeks, we'll be in great shape."

He looked excited. He was shifting around, playing with the cinders. He pulled the hood of his parka down, took his hat off, and came upstairs. "I just hope we don't have another whiteout," he said. "That was enough excitement for one lifetime. People are still asking me about it. What can I say? I'm glad the pilot remembered to climb."

FIFTEEN

The Flume

It must have been the combination of the woman's voice and the absolutely pleasurable feeling of direct sunlight falling through the freezing night air that set me to dreaming. I was imagining the smell of flowers and cut grass, the sight of women in flowing dresses passing along crowded streets, the sound of water splashing on patterned marble, the wheeling of stars over a dark tree-lined ridge—things I had not known for a long time.

▼ ▼ ▼

IN A PARKA AND WINDPANTS it is possible to sleep comfortably anywhere. The small, second-floor sun deck of the Hotel California, with its sanded wooden planks and its perfectly horizontal lines, was a far more inviting surface than the sands and ventifacts and deceptive hollows of the valley. On waking there, I would usually find myself pitched against the tent or perhaps turned around ninety degrees from my original position. So when I started to doze off on the sunlit deck, I only half resisted it, knowing that a night there would be more comfortable than any I had spent in the field. I stretched out fully on the wooden floor.

A gentle calm settled over McMurdo. The streets were nearly vacant. From the doorway of the Biolab I heard the voice of a woman laughing. The helicopters stood motionless on the pad, their blades neatly sacked in canvas, roped forward, and tied down. The heavy equipment, which from early morning until late night, crawls and chugs and churns its way through the dusty streets, was silent. Across McMurdo Sound, ice pools caught the evening light like great telescopic mirrors and focused it in brilliant points and flashes into the lucent air. From the sea beyond Hut Point there was a huge fog bank moving toward town. The flags of the signatory nations, which stood in a bright semicircle around the marble bust of Admiral Byrd, barely moved in the light breeze. The midnight sun felt warm against my face.

It was six in the morning when I awoke. I hadn't moved all night and my face was still turned up toward the sun, which by now had half-circuited the horizon at its cold summer angle. I walked back to the room a bit stiff from lying on the boards. I was rotating my head around on my shoulders to assure myself that I hadn't turned to stone. When I opened the door, Varner was sitting in the artificially darkened room over by the window. He was fidgeting with his matches, trying to light a cigarette. There were nervous scratching sounds. Little flickers of light danced in the air. Finally a wooden match caught and burned steady in the darkness. The blue light played on the surface of the window shade, then died out. I could see the red glow of his cigarette moving around.

"Where were you?" he asked. "We looked for you everywhere—the lab, the dining hall, over at the officers' club." He was wide awake. He looked as though he had been sitting there for hours.

"I fell asleep on the deck," I said. "I dreamed a lot. I dreamed I was home in bed, in Ohio, dreaming I was here. Wanda was saying, 'No, it's okay. You're really at home. Don't worry.' It was strange."

"Are you ready for this week?" I asked him. "I think we're in good shape."

"I hope so," he said. "There's a lot riding on this for me."

"The whole project's riding on it," I said. He was nodding his head, staring straight at the red glow before him.

"Maybe we should get packed," I said. "Get some food. It's a mess out there, but it could clear. They might fly at nine."

"It's always a mess out there," he said. "But you're probably right."

Mike was right in his estimates. In addition to all of our meters and bottles and field equipment, there were the sleds and the drill bits and motors and all the countless camping supplies: tents and stoves and freeze-dried foods, all of the carefully chosen stores from the Berg Field Center—beef stew and spaghetti and peanut butter and cheese and freshly baked bread—enough for the three weeks we would be back in the field.

We loaded all of this into the truck—first at the BFC, then at the Biolab. Then we headed down to the pad. There were two helicopters waiting, and two crews. The sergeants were out, checking the machines, ducking under them, looking at the skids, rubbing their hands over the blades, which drooped like the fronds of giant ferns across the cindered deck. The pilots, in their dashing flight suits, their white turtlenecks and silver wings, leaned near the hangar door talking, coolly oblivious to the commotion around them. The BFC truck rumbled to a stop between the two Hueys.

Varner and Walt were moving carefully down the cinder hill to the pad, carrying between them a large, V-shaped object. Occasionally it would catch a gust of wind and shift them a little to one side, so that it looked as if they were staggering. When they got onto flat land they put it down and rested. Varner bent nearly in half to shield himself against the wind. He lit a cigarette, straightened up again, and started to pace. He was pointing at the flume, at the stilling well, saying something to Walt. He looked agitated.

It took us half an hour to fill the two cargo bays. We loaded the flume first, wedged it in behind the pilot's seat. In the empty spaces, in the arms of the V and in the stilling well, we stuffed sleeping bags and boxes of food and cartons of plastic bottles. Varner and Walt strapped themselves into their seats and Mike and I walked to the other Huey, where we had stored chemicals and fuels and batteries. We were ready to leave. The sergeant closed the doors, and the engines of both Hueys came alive. Thick jets of white condensate rolled from the exhaust and opened upward into the air. The mountains were faint shadows in the distance. The sun was a thin disk, white and cold and far away.

▾ ▾ ▾

The flight into the mouth of the valleys was uneventful. Mists drifted by the windows but remained tenuous and unthreatening. The visibility was good. We flew into New Harbor and up the north shore of Lake Fryxell, over the green wooden hut the Kiwis had built on the east side of the Canada Glacier, and finally to the rocky shore of Lake Hoare. This sector of Taylor Valley is defined on the east by the massive presence of the Canada Glacier, whose front rises abruptly in columns of fluted ice some seventy feet above the valley floor.

There is a small stream that tracks the north wall of the Canada Glacier. Near the ice margin the stream has cut a narrow little canyon perhaps three feet at its deepest, but steep-walled and dramatic. In its entire length, from where it forms up on the northeast margin of the ice, to where it moves away from the glacier wall and sets out on its own course toward Lake Hoare, the stream is only three miles long. But from the maps of the valley it was clear that this was the major stream feeding the lake, the major source of dissolved salts. We named it after my daughter. We called it Dana Creek.

▼ ▼ ▼

Once on the ground, we organized ourselves quickly. The first order of business was to get the flume in before the stream rose. We extricated it from the pile of gear that lay around and on top of it, and carried it over to the streambed. There were gray, striated snow clouds hanging low over the mountains. The valley, in this weak wash of dirty light, appeared even narrower than it actually was, even more confined. We set the flume into the sandy creek bottom, turning it so that the V opened up toward the glacier as though its arms were welcoming the spring flood that we knew would come. The notch of the V, through whose narrow opening the waters would pour, faced downstream to the lake. We wiggled the plywood frame into the sand, working it a few inches below the surface so that it would hold.

Varner stepped back and looked at it. He lit a cigarette. He was standing there, cocking his head a little. "This looks right," he said. "I think this looks right." He was shaking his head up and down as though he were admiring a piece of sculpture. We were all staring at it, at the trickle of water that was beginning to twist like a silver thread over the plywood base, moving slowly through the V on its way to Lake Hoare. "It's beautiful," he said. "I just hope it holds when the flood comes. That's been my fear all along. One of them, anyway."

I had no doubts. I knew it had to work. We were too close. The flume, sitting in the midst of the channel, seemed absolutely right, as though the two had been designed with each other in mind. Instead, my thoughts were racing ahead to what we would soon know about the chemical weathering of these rocks, about the fate of the metals, about the origin of this lake.

The stream at my feet was beginning to rise. As the water from the melting glacier fanned out and lapped the stone, a small part of Earth dissolved. Even the atmosphere responds: some of its carbon dioxide enters the stream and reacts there, reacts with this mineral or that, is converted in the weathering process to bicarbonate ions. Robert Garrels, the great geochemist, estimates that the annual weathering of minerals at the Earth's surface removes enough carbon dioxide to deplete the entire atmospheric reservoir in two thousand years. All of this is going on at my feet, on a bitter, overcast day, in a tiny triangle of sand and lake and stream, in the shadow of a creaking glacier, at the ends of the Earth. But it is going on everywhere, in every clump of spartina on every tangled bank, on the scaling surface of every monument, in the curve of every shell, in every cloud—wherever stone and water and air meet, wherever the elements join and mingle in their journeys.

If the Onyx River at Lake Vanda was any measure of things, it would happen here slowly. It would happen the way it did on the early Earth, in the Precambrian dawn of life, before there were land plants, before there were the bright fuchsias and oranges dusted on the fields, before there was opulent decay, before there were all of the organic acids released for chelation, for prying loose metals from naked stone. It has been estimated that the rate of chemical weathering is increased fourfold by the presence of land plants. And so they conspire to bring the continents down, to turn them to runoff, to quicken the loss. Everywhere but here.

I was standing on the banks of Dana Creek. It was swelling and swelling as the clouds scattered in the east, as the pale sunlight began to fall over the blue ice. Varner had already installed the float and the strip chart recorder, with its eight-day, weight-driven clock. The yellow float, which had dangled like a bright pendulum in the well, which had touched only cold air, was now rising with the creek. You could see the pen move against the chart, grope its way upward among the red lines, as though driven by a spirit. As the sky cleared, as the light striking the glacier intensified, the pen climbed more steeply. You could see the face of the day written and remembered in that single line.

Varner was chain-smoking now, tamping out the butts on the back of his gloved hand and depositing them in the pocket of his parka. "Jesus Christ," he muttered, not really believing it, "the damn thing is working. It's working!" He was pacing back and forth along the banks of the stream. You could hear his boots crunching in the sand.

He stopped abruptly and sank to his knees. It was as though he had been punched in the stomach. He was digging in the wet sand at the base of the flume. He had taken off his gloves and was scooping up sand in his bare hands. "Oh, no," he said, "it's leaking, there's water coming up from below. It had to happen. It had to."

We all went down on our knees and tried to cement the flume. But as soon as we patted a handful of sand against the leak, a new leak would spring up nearby. When we patched that, our other repairs would disintegrate, be flooded away to the side. We were losing too much water. Maybe forty percent of the flow was seeping under the frame. And still the stream was rising.

Mike looked at me and said, "I brought some sandbags. I don't know how many. Should I get them?"

"Get them," I said, "they might help." He ran over to the tents, opened a large footlocker, and flung out a pile of burlap. Then he came running back to the flume, his arms loaded down with burlap. The flume was beginning to shift on its base. It lurched through a small angle. It seemed as though it wanted to float away.

Varner and Walt grabbed the shovels. Mike and I stood there with the bags opened before us. When they were filled, we tied them with twine and hefted them against the front of the flume. Our work became rhythmical. The sand hissed out a kind of tune as the shovels struck it, as it slid off the metal and into the bags. We were filling and pitching, filling and pitching as if we had been doing this for years, as if we were a hired crew that specialized in worldwide leak control. Varner was supervising the placement of the bags. "Over here," he shouted, "it's still leaking under the front. This baby is going to hold."

▼ ▼ ▼

I'm not sure how long it took. Three hours. Five hours. Maybe more. But the creek had risen to where it was halfway up the walls of the flume. At one point there was a pulse of high water that must have been stored behind an ice dam in the glacier. It came at us all at once like a small wave, rolling rocks the size of baseballs before it. Then, just as swiftly as it rose, the creek began to subside. We had caught nearly all of these changes. You could see it on the graph, the way the line climbed steeply, then fell a little with the leak, then rose again as we bagged the flume, then shot up abruptly with the pulse. The hydrograph was a written memory of the hours, of the rise and surge of water in a stream, of the sun against the ice. For Dana Creek, perhaps it was the first memory. A little handful of time on Varner's chart.

We sat on the smooth sand, looking down at the stream. You could hardly see the plywood of the flume. There were only the breadlike loaves of burlap stacked around it, even on top of it, to give it weight. The pen moved with a jittery hand across the face of the chart as the yellow float, and ultimately the water in the stilling well, directed it. The conductivity meter stood atop the flume, its probe dipped into the stream, its red needle barely lifted off the zero. There was a pH meter next to it that read 7.0, neither acidic nor basic, as though the waters of the stream had been deftly titrated to pure neutrality by the acidic carbon dioxide of the air and the basic salts of the valley. Varner was inhaling on his cigarette, holding the smoke in for a long time, then releasing it with a calm and pensive exhalation. I could imagine him sitting there for months, in that same position, watching the pen skitter on the page, watching the water dance over the wood on its way to Lake Hoare.

Walt and Mike were quietly talking about their projects. Walt said, "You know, this is the first time I've believed I would be able to get the data. First there was the problem with phosphorus. Then the streams. But maybe we have it now." Mike agreed. "With Varner's flume," he said, "I think I can write the history of this lake. I think I have a thesis."

It felt very late. I was tired and more relaxed than I had been in weeks. I told them I needed some sleep.

The Scott tent into which I crawled was a tall green tetrahedron. Its tip was eight feet above the sandy beach. I had rocked it down with a few boulders that I had gathered earlier from the slopes of the Asgards. The light inside was dim enough that I could sleep without winding a black band over my eyes, but not so dim that I couldn't read or write. There was enough room for a wooden crate that I used as a kind of nightstand for a few books and for my journal. I placed a foam pad over the canvas floor and spread my sleeping bag on it. I removed my parka and balled it into a soft pillow. Through the porthole at my feet I could just glimpse the ice of Lake Hoare. Behind me I could hear the sounds of the Canada, its creaking and shifting and moaning, the tinkling of its falling ice. At times I thought I heard the wind in the towering ice as though it were an instrument.

I did not fall asleep for several hours, but lay there listening to the lake and the glacier and the stream, whose flow I could barely hear. Just a far-off babbling sound, as though it were breaking over large stones. I made notes in my journal so that 1 could pass them on to my daughters. Soon the wind was roaring, and I could barely hear the stream as it flowed through the flume. I turned over and paged through my journal. I was looking at something I had written about the gallery. One entry said, "What about Northrop's Stream?"

"Northrop's Stream" was like Eddington's table, one of those images found in an essay read long ago that had lodged itself forever in my mind, that could be teased out and set free by almost anything: a chance encounter in a garden; the right smell carried on just the right breeze; the sound of water barely audible above the wind. The essay began with a poet on the banks of the stream. For the poet, the stream is a "babbling brook," a congeries of sound and light, rippling surface flashing into brilliance, onyx-black water patched and swirling and ever changing, subducting itself under the stones. It is a lovely confusion of tone and motion and sound coming all at once and then disappearing, as the philosopher F.S.C. Northrop says, "into an immediately felt vagueness at the periphery of our consciousness." It is the stream of Heraclitus that the poet sees, the stream that one can never stand in twice, the stream that one comes to know only in the stillness of memory, as a trace, as something tasted on the tongue.

The scientist on the bank "sees" something else. The scientist sees the transit of molecules, the broken flight of water as it arcs and jostles and collides, as it aggregates and disaggregates a billion times. Or perhaps the scientist sees velocities and weathering rates, laminar and nonlaminar flows, the constant little exchanges of heat that accompany every movement of every parcel of water through every shade and elevation on its way to the sea.

According to Northrop, poetry and art are concerned with immediately apprehensible material, with what we experience with our senses, with the emotional content and vividness of that experience. Science, on the other hand, deals ultimately, at its theoretic and deepest levels, with what is not observable to the senses, with the abstractions of electrons and protons and ions and flickering clusters, with plates and waves and quanta, with an imagined world beyond touch and sight and smell.

How odd this view is when compared with our customary reckoning that science is somehow nothing more than a body of facts. If it is fact and description that are required, then poetry and painting would be the better science.

Northrop had distinguished the two—art and poetry on this hand and science on that. But it was all poetry, the distant table and the little stream, all "geopoetry," just as Harry Hess had said.

As I listened to the stream breaking over the stones beyond the tent, listened to it "babble" just above the nearing wind, and as I thought about it in all of its ways—its spun silver and its speech so readily apparent, its nanosecond encounters and existences, its calcium stolen from stone—I realized I could not tell where one account left off and the other began, so real and palpable were they both.

Eddington's table, flat and seemingly lifeless in the gardens at Christchurch, stands beneath the abundant monkey puzzle tree and the thin laburnum. But what lay beneath its flatness? What of the vast energies and structures it contained? What of the forces strong and subtle that held it in place, that twined its cellulose into cords of wood, that made it endure? Once it was a tree and the tree had come from air, had almost condensed from it, and itself would one day return to air, as if by magic, but with a certainty you could just count on. Wind and water would wear it away, fire would set it free of its bonds. In the soul of matter was the memory of change, and in the end nothing endured but that.

I was still wide awake. There was noise everywhere. The up-valley wind sounded like a long Midwestern freight rumbling by, and the tent walls were beginning to flap wildly. The stream, to which I had been listening so intently, effortlessly folded its sound into the wind. Then, suddenly, the Scott tent, all eighty pounds of it, began to shake and uproot itself. I heard the rocks I had poured on the front slide and spill onto the ground. Suddenly the tent lifted from over my head and was flying up toward the glacier. I was lying there, faceup on the canvas floor, staring open-eyed into gray cloud. I rolled over and watched the tent go. It landed and bounced and glanced off a boulder, spun around it as though it were breaking a tackle, and continued its flight toward the Canada. Then it lodged against some calved blocks of fallen ice. It stopped and lay there, flapping, like a hurt bird.

I was in the open air of the valley. The wind rushed over me. I sat up and threw my parka on, pulled my hat down over my ears. I turned toward the glacier again, and saw the tent still caught there on the ice. I shook my head. I had to laugh a little. All of a sudden I was sitting in the middle of nothing, on the shores of a lake that had dropped out of

the Ice Ages, that had a plaintive and prehuman feel to it. I raised the
hood of my parka and folded my arms around myself. I could see the
lake, with its ragged ice, running far up valley, tapering as it went to
the width of a stream. With the tent gone, it seemed quiet, only the
moaning wind.

Off to my left, I heard an orange tent move. I saw a bright red
balaclava poke through the hole and look around. It was Varner. He
crawled out and walked over. "What happened?" he asked. "Where's
your tent?" He was speaking over the wind. I pointed up to the Canada.
He turned around and saw it. "Jesus," he said. "You had it rocked down,
too." "Not well enough," I said. "It's really wild here tonight."

He sat beside me on the sand. He was fiddling with a cigarette,
pretending it was lit, that he was smoking it. "Just before we took off,"
he said, "I recalculated the depth of the stilling well. 1 thought I had
made a mistake. On the way out I was convinced I had it wrong. But
with the first flow, when I saw the water lift the float, I knew it was
right. I had it right all along." I didn't need to worry any longer. He
was consumed with it now.

There were gusts of wind blowing through the flume. It was whipping sand around, glancing it off the plywood. There was no water.
The ice had all sublimed. I thought I could hear the frame creak on its
base. Varner was saying, "I never really believed I could do this. Not
after everything the hydrologists back home had said. So I convinced
myself I didn't care. Now I just want to see the numbers come off the
chart. I want to see what makes it all tick. I have some ideas. I think
maybe it's the way the sun strikes the ice."

Varner was looking at the flume, pulling the biting air through the
unlit cigarette. "I think this will give Walt what he needs now for the
nutrients, and Mike too. I think this baby is going to hold, if the wind
doesn't blow it down."

I told him I was amazed at how well it had worked, how glad I was
he had stayed. "No one has ever done this before," he said. "I feel like
John Wesley Powell on the Colorado." He laughed. I slapped him on
the back. "I think the place has gotten to you," I said. "We're going to
have to pull you away."

"Sometimes, you know, I think I'd rather be back in Akron.
Sometimes I wish I had grabbed that plane a few weeks ago. But not
tonight. Tonight I feel like I should be right here, right by this stream.
It's all come down to this."

He went off to his tent, crawled through the small porthole, and was gone. I sat there for a while looking over the lake, imagining how it must be in winter, with the raw wind and the darkness and cutting sand everywhere. With the stars rolling up there above the ice.

When I glanced down at Lake Hoare, I saw something moving. It appeared to be upright, shuffling between tabletops of crystalline ice. Its gait had a shambling quality to it. I rubbed my eyes a few times, thinking it was a kind of light trick, an Antarctic mirage. When I focused on it again, it had moved closer. I could see it more clearly now. It was waddling from side to side, brushing awkwardly against the sculpted ice. As it neared the shore, it stopped a second and looked down at its feet. I had to pinch myself. It was a penguin, a small Adelie.

In the middle of Taylor Valley, I said to myself. *How did it get here? It's sixty miles from the nearest rookery. There's not a shred of food. It will die in this place. There's nothing here for it at all.* I rooted through a footlocker and pulled out some lobster tails that Walt had gotten from the BFC. I walked down to the ice. The wind was blowing so hard that she had to stop and lower her beak into her breast. I thought the wind would topple her. I laid the lobster at her feet. She stretched her flippers out and shook them up and down a little as a warning. "Stop," they said. "Come no farther." She was swaying from side to side. There was no noise. She was looking up at me, her eyes wide and as black as onyx.

I retreated, backing off the ice and onto the shore. I walked up to where the tent had been and sat on my sleeping bag. She was lying on her stomach near the lobster tail, pecking at it with her beak. I called Varner several times. But there was no answer. He had fallen back to sleep. I crawled into my sleeping bag, pulled it up around my head. The tent was still fluttering up by the glacier. I felt tired, but so alive. More so even than I had felt on the deck, dreaming, a lifetime ago. I guessed I would see the Adelie again in the morning when I awoke. There was no easy way out of here. It felt good that there were five of us now.

SIXTEEN

The Cone of Erebus

It is no wonder, then, that geochemists trouble themselves over these seemingly insignificant elements of the crust, these elements of dead stone. In the pageantry of Earth, they are as vital as water itself. No wonder that geochemists have marveled at how unusual this Earth is, how right its long history, how one-of-a-kind in all the universe it may be. How one-of a-kind we ourselves may be.

▼ ▼ ▼

WE STAYED AT HOARE FOR A WEEK. There were days when the lake turned to pure crystal. Other mornings it would be dusted with dry snow, and we could see the tracks of the penguin wandering and looping around and back on themselves. They would disappear far down valley, past the flag that someone had planted in the ice years ago. There were so many glistening mesas and channels and heaps of sand on the long surface of Lake Hoare that it would have been pure happenstance had we seen her. Late in the week, we saw what we thought were fresh tracks. We followed them from mid-lake onto the shore. They stopped suddenly and clustered in nervous little scratchings around the lobster tail, then moved up toward the Canada Glacier, to where we could no longer see them. She appeared to be walking out of the valley, down toward the sea.

The flow of the stream had become almost rhythmic as its rise and fall tracked the movement of the sun. In the mornings, when we awoke, we would find only a trickle running along the base of the flume. But as the day wore on, the water would rise and we could hear it like a mountain stream, babbling and gurgling and roaring through the gorge it had cut in the tills. From out on the lake it sounded at times like a distant highway heard from a motel room, like a long, steady line of traffic up on the interstate.

I had rarely seen Varner so excited as he was now. From my occasional visits to his classroom over the years, I knew him to be a great teacher who could hold an audience spellbound with his stories. I remembered him once standing there with his arm outstretched, moving his hand slowly from side to side, as he recounted the tale of the cathedral lamp, of Galileo's pendulum sweeping silently above the turning Earth, of the turning Earth itself, that had earned Galileo such opprobrium in those days. The students seemed hypnotized by it all. How could Galileo know this, they wondered; how, against all the evidence of the senses, against all the received ecclesiastical wisdom, could he know that the Earth turned? Varner had re-created the times for them, had transported them to the very moment of discovery, of creation, when Galileo knew in his gut that the Earth really did move.

Later I saw him do much the same thing with Newton. He was holding a piece of Icelandic spar, a calcite prism, through which he was filtering light, watching it spread in the darkness into curtains of color along the walls. "Light can be broken down into parts," he said, "just like matter. It's atomistic, in a sense." Why, he wanted to know. How could they explain this? What must light be like, for this to happen? What must the prism be like? He was teasing them with questions. Smiling. Urging their explanations closer and closer to the truth. It was as though he were reliving the moment of creation himself, and they with him. You knew why he had become a teacher.

There was that same intensity and excitement as he worked on the stream. He arose early, got to bed later than the rest of us. He was seeing patterns in the way the stream flowed, in the way it responded to the Sun, to the light thrown on the glacier walls. The stream had become a kind of armature for him in the larger workings of the universe, and he was trying to descry its movements, to disentangle the true causes of its flow from those that were merely apparent. How important, for example, was the temperature, the time of day, the angle the sunlight made with the north face of the Canada? He was pacing the banks of the stream, knowing without a doubt that he was on the verge of something important. It would all come together soon in the tracks that the pen was making on the chart. Finally, I think, Varner was doing what he really came to Antarctica to do. He was doing science.

One evening after the sampling was done, we put on our twelve-point crampons and climbed to the top of the glacier. It was not difficult, though there were times when we had to wedge ourselves into chimneys of hard blue ice and pull ourselves onto ledges. At the top, a landscape of silver rivers and winking lakes and deep yawning crevasses awaited us. These were the headwaters of the cascades that we had seen from our camp, that rushed over the glacier's snout, maybe the purest waters in the world. On our knees, we reached with cupped hands into the deep, narrow ice gorges that the streams had carved, and drank from them. *Desert*, I had to remind myself. *We are in a desert.* We were high above the floor of Taylor Valley, and we could hear the wind singing among the columns of ice we had just scaled.

We were beginning to understand Lake Hoare. Only a week's data had come in, but already we were doing back-of-the-envelope calculations in the tents at night, punching up flow rates and ionic concentrations, estimating fluxes. From the early numbers, it appeared that Hoare was very different from Vanda. Its waters were only weakly stratified, its dilute brine was of the sodium bicarbonate type, like certain lakes in the African Rift Valley and in the Basin and Range province of North America. There was no evidence of the mysterious, calcium-rich groundwater that had so bedeviled our efforts at Vanda.

The hydrology of the lake was becoming clearer to us as well. There were only two streams: Dana Creek and a very small and intermittent inflow from the Suess Glacier. Much of the water came directly from the Canada Glacier; we could see that from camp. Behind us, on a sunny afternoon, there were waterfalls plunging over the face of the glacier, pouring into frigid ice pools and onto the sandy beach. As it flowed over the sand it warmed slightly, to about four degrees Centigrade. When it entered Lake Hoare, it sank beneath the less dense waters of the surface until it reached a constant density. Then it spread out as a thinning film, all the while imbibing ions from above and below, as it came quickly to equilibrium with the lake.

For a while it all seemed too simple, too straightforward. I was becoming nervous. I remembered something my father was fond of saying about life. It was one of his favorite expressions. "Look out for the knuckleball," he would warn me when I got too cocky. "Life's been pitching them all down the middle to you. That's when you just know the knuckleball's coming."

The world threw mostly fastballs—sometimes down the middle, most of the time a little low and outside, or high and inside, a bit off center. If your eye was good and your timing was right, you could hit them out of the park. But then, when you weren't expecting it, in it came. It was slow and big. It looked as ripe as a mango. You cocked your bat and got ready to drive it into the stratosphere. But then you noticed it was doing funny things; it was dancing and floating and shifting around in the air. It moved like a butterfly, and you just couldn't tell where it would go next. You were squinting and saying, "What the hell!" When you swung, you connected only with the air. *Whoosh*. A billion molecules of oxygen and nitrogen whirled across the surfaces of the bat. There were invisible vortices swirling all around you. The ball, when you turned to look, was as white as Erebus against the catcher's mitt. It was the knuckleball. Your confidence just evaporated.

Nature has its knuckleballs, too. Philosophers of science call them anomalies, but whatever name or metaphor you prefer, they are the things that keep you off balance: the faint glow of a cathode ray tube; the recoiling of alpha particles from a thin sheet of gold; the disappearance of magnesium, whole unaccounted mountains of it, from the sea; the calcium chloride brine of Vanda. Just when you think you have it right, when you are swinging from your heels, in comes the "knuckler."

It was like that on Hoare. We thought the water and the salts came mostly from the stream we were measuring, Dana Creek. The Canada Glacier, for all its looming and spectacular presence, was, in our first model, a minor force. But we were wrong. One night, after the wind had died off, I went out to the deep hole, about half a mile west of the glacier. Blue sky. The castellated mountains. Silence. I could hear my blood flowing again. I lowered the probe of the conductivity meter slowly through the water column. As it sank, the red needle shifted across the face of the meter. The conductivity was increasing with depth, just as it had done numerous times before. Then down a little more into the still lake. Suddenly the conductivity dropped. *Instrument problem*, I thought. *The battery's going*. I brought it up a little. The conductivity rose again. Same value I had gotten on the way down. I went down again. Low conductivity. It wasn't the instrument. Something was going on down there.

I pulled the sampler off the sled, opened it, set the trigger, and sent it through the ice hole. I lowered it to where the anomalous water lay; then I shot a messenger down the line. I pulled the sampler up, poured the water into plastic bottles. Using large volumes, I did the calcium and chloride titrations right on the ice and computed their molarity as soon as I was through. I looked at my notebook, ran my finger down a long column of numbers to where it intersected an entry marked GLACIAL ICE. I looked over to the column labeled CALCIUM. The value recorded there was the same, within a few percent, of the value I had just obtained from the lake. I did the same for chloride. The agreement was even better.

There was only one explanation. Far beneath the surface of the lake, where liquid water lapped against pure ice, the Canada Glacier was melting. I had heard reports from the biologist George Simmons and his divers that they had once seen streaks of water on the glacier's face, and little tendrils and freshets shooting deep into the lake. I had thought of this as a rare event and as nothing important. But there I was, half a mile away from the submerged snout, intercepting what appeared to be a river, a frigid Gulf Stream flowing into the midst of Hoare.

I put my notebook down, closed it, sat on the sled, and rested. It had seemed so simple a few minutes ago: just two streams, a lake and maybe a little input, mostly overland and visible, from the glacier. But that was wrong. The Canada was important. There was water and there were ions—ions blown in from the sea, perhaps, lost for a while on the land, trapped in a valley, in a glacier. They would have to be taken seriously. We had some modeling to do.

The sun was so warm that I couldn't resist stretching out on the sled. It was three in the morning and it felt like summer. I was watching a ball come toward me. It was as large as a mango. I could see its seams against the blue sky. It was dancing. I reached to hit it, but it dropped away. The air rushed across my bat. I heard my father's voice. He was saying, "Watch those knucklers. They'll get you every time." I opened my eyes wide and turned to where his voice had been. I was looking straight at the Kukri Hills. They were brown and empty, except for where the spring snow clung in hard little patches to the slopes.

▼ ▼ ▼

You could hear the Huey as soon as it entered the valley. At first it sounded like a drill, faint and indistinct and far away, but then you could make out the *thunk* of the blades as it pushed along the Commonwealth. When it landed, the pilots didn't even bother to shut down. Mike and I loaded our supplies under the rotors, ducking through the propwash with our eyes closed. We said good-bye to Varner and Walt. There were handshakes and thumbs-up and genial kidding shouted over the engine noise, and then we climbed aboard and were off around the glacier's north face. The plan was to sample Lake Fryxell over the next few days, and to study whatever streams might be flowing in the lower Taylor Valley. Then I wanted to return to Vanda to help Tim and Dr. Yu with a new array of traps. We had all planned to spend Christmas at Vanda.

Even though the distance between Hoare and Fryxell is only a few miles, and even though they share the Canada Glacier as a common terminus and source of water, the sense of space is very different from one to the other. East of the Canada Glacier the landscape stretches and widens and opens onto vistas of the Ross Sea. The sky asserts its presence more. The feeling, in contrast to that at Hoare, is one of connectedness; of having the whole varied world somehow present and within view. I have heard the Apollo astronauts talk of seeing the Earth from space, of taking it in all at once, of being deeply united with it for the first time. Here the union, though occasioned less by visual cues, was no less strongly sensed. At Fryxell I sometimes felt as though the world were resting on my tongue, as though it were dissolving like a host.

The pilots put us in near the green "Kiwi Hut," high on a rocky moraine overlooking the lake. We were two hundred feet above its surface, above the indentations of its coastline and its small, dark islands and deltas. What was most unusual about this location was the view that it afforded of the volcano. From the Kiwi Hut, high on its perch above the Taylor Valley, it was possible to see the cone of Mount Erebus. In the evening, when the sunlight came from the west, its flanks gleamed. There were rivers of light flowing down its side, and a flocculent plume of vapor carrying downwind from its crater. Though it was nearly sixty miles away, its sheer size and the clarity of the air made it seem as much a part of this landscape as the streams and the glaciers.

We seemed to be moving very slowly in our work. But it was often like this after we had transferred from one camp to another. There was a slight flagging of energy that attended these moves, a growing reluctance to begin all the hauling and drilling and casting for samples again—an impatience, perhaps, with the long shadows that fell between our arrival on station and the first data.

With everything in order, we went into the hut to make ourselves a drink and a sandwich before we hiked down to the lake. Mike was holding a white cup in his hands. You could smell the sweet chocolate in the air. He had taken off his parka and was sitting across from me in his checkered wool shirt. His dark beard was tangled around his mouth, and his hair stood out in wild tufts from his head. He closed his eyes as he tasted the hot drink, and his face lingered for what seemed like a minute in the rising steam. His head was sinking slowly toward the cup.

Mike had organized everything. He had gotten us from one valley to the next without a hitch, without forgetting so much as a cotter pin or a spare bolt. He had done more than his share of the hauling and drilling. His project on the origin and evolution of the lakes was beginning to take shape, especially now that the preliminary results on Miers and Hoare had suggested that these were very different from Vanda, and even different from one another. The problem for Mike was becoming clear: How do you account for the fact that the chemistry of these lakes—all located in roughly similar glacial terrain, all ice-covered and fed by a small number of streams, all depauperate in their biology—is so remarkably different from one to another. Miers was fresh water, calcium and bicarbonate; Vanda was brine, calcium and chloride. Hoare lay between these two in its overall saltiness, a kind of compromise; it was sodium and bicarbonate. And Fryxell we didn't know yet. We couldn't even guess. We wanted to get the data and begin to sort everything out. "The grand unified field theory of Antarctic lakes," Mike proclaimed, laughing. "We're nearly there."

The Optimus stove burned with its low, throaty roar, and the room was becoming warm and heavy. Mike had laid his cup aside and was sitting in the chair with his eyes closed. "Why don't you take a nap," I said. "Get some good Z's. We have time." He nodded and moved over to the lower bunk, stretched out on his back, and began to snore lightly. I poured some hot water into a porcelain cup the New Zealanders had left, mixed in some hot chocolate, turned off the stove, put on my parka, and walked outside.

I had begun to feel drowsy in there, but the cold air soon awakened me. There was no wind. I could hear my boots crunching against the sand and rocks as I moved. I could hear the stream babbling out of the ice grotto. From where I stood, I could see the stream and the lake and the white cone of Erebus far across McMurdo Sound. For a thousand square miles beyond the hut there was not a single living thing.

It was odd, standing there, to think of how the volcano and the stream and the shells I had collected here and elsewhere, and the very ground on which we walked back in Ohio were all linked as if by some invisible bond. Paul Klee said, "Art does not reproduce the visible; rather it makes visible." What was made visible by geochemistry and made immediate by the juxtaposition of the volcano and the stream and the glacier—all within the compass of my vision on the barren hill above Fryxell—was the great cycle of carbon and the cycles that turned with it. Looking at the volcano, I could trace its plume, its carbon newly liberated into the world of light, to places and times that are far away, to springs and lakes and rivers, to the deep snow, somehow remembered, that fell in linen shawls upon my small landscape many years ago.

The snow that fell that night, that lay heavy on the rooftops above the silent attics of the town, was not water only. It was water and bits of earth and traces of gas. A wisp of carbon dioxide had stolen into the water before it condensed and had branched and turned, six-sided, to snow. The molecules of gas joined with the molecules of water to form carbonic acid, a "weak" acid, but, to the geologist, the very acid that etches mountains from the Earth, that limns canyons on its face.

When this remembered snow disappeared, it trickled over the flat-worn bricks, swelled in the gutters, and rushed to the streams. At the mouth of the creek it joined the river and flowed westward. Into the flow crept ions and solutes that had come from rock—from solid calcium carbonate and calcium silicate. In the sea, bicarbonate and calcium joined, not at once but in the fullness of time, to become part of a shell, perhaps a shell tossed through an August rain onto a barren shore; or perhaps the spiral of a coccolith in the spring blooms, when the sea turned to cloud.

And the shell, the calcium carbonate, became part of the sediments. And the ocean sediments sank into dark trenches and seeped under continents into the fiery mantle of the Earth. And there, in the deep magmatic chambers, it reacted with silica, was transformed again to carbon dioxide. Through the vents of volcanoes, through the narrow throats of the crust, the mantle communicated with the air. The volcano's exhalation, its breath, came from far below the mountains, from orange caverns that were on fire with molten rock. This breath was part Earth and part sea, part stone and part shell. It was the mountain's respiring. It was dense with carbon dioxide.

One of the great tricks of life has been to learn to sequester carbon. It has done this by stacking it in brilliant cliffs along the Dorset coast; by burying it, hundreds of feet thick, beneath the hills and farmlands of Ohio; by interring it in caves and quarries and corals, entombing it in mountains of marble, turning it into stalactites and stalagmites, and heaping it like fallen snow in the silent sediments of the sea. Enlisted in this great warehousing of carbon, in massifs of marble and chalk, is calcium. For every carbon atom woven into the proteins and fats and carbohydrates of living things, there are 62,000 atoms of carbon in limestone alone. Most of that is the remains of marine organisms. To a lesser but still important degree, the same is true of magnesium. The Earth's repository of dolomite, rock consisting of mostly calcium and magnesium carbonate, holds 45,000 atoms of carbon for every carbon atom in the biosphere.

Why is the storage of carbon in the bones of the crust so important? Carbon dioxide is a greenhouse gas; it absorbs infrared radiation tossed off by the surface of the Earth—energy that otherwise would escape into the cold of space. With insufficient carbon dioxide, the Earth would be a frozen sphere, like Mars. Its seas would be capped with ice like Miers and Vanda and Fryxell and Hoare, or perhaps they would be frozen all the way to the bottom, like Victoria. With too much carbon dioxide, the Earth would be an inferno, like Venus, whose surface temperature is four hundred degrees higher than ours. Its seas would boil and roll as dense clouds across the land. That these things do not happen is a testament to the way in which carbon has been husbanded and stored and partitioned between rock and atmosphere.

In the beginning, when the sun was dimmer, there was just enough carbon dioxide and other greenhouse molecules to keep the Earth blanketed and warm, to keep the seas liquid. With liquid seas and all that entails—cloud formation and rain and the dissolution of rock—the great cycles, powered by a weak sun, could begin to turn.

The first life began beneath the waves in waters that were anoxic, reeking of sulfur, like the brine of Lake Vanda. More than three and a half billion years ago the Earth produced its first chlorophyll—a ringed molecule of nitrogen and carbon, in the center of which sits a single atom of magnesium—which allowed photosynthesis to occur. With photosynthesis, everything changed. Carbon dioxide and water were transformed into living cells, into organic matter, and, as a byproduct, gaseous oxygen was released. The carbon dioxide of the atmosphere was gradually drawn down, while at the same time the level of atmospheric oxygen rose. When the carbon cycle began to turn, the pressure of carbon dioxide was about sixty atmospheres; today it is about three hundredths of an atmosphere. Over the Earth's history, the concentration of carbon in the atmosphere has been reduced by a factor of two thousand.

Gradually the carbon of the ancient atmosphere has been stored: in the kerogen and ooze of the sediments; in the bicarbonate of pelagic seas; in seams of bitumen and anthracite; in pools of oil; and in humus and caliche. But mostly it has been stored in shells. Had it not been this way, had the Earth not hoarded its carbon in hidden troves, the surface waters would have long ago turned to steam.

Two geochemists who have studied the global carbon cycle, Robert Berner and Anthony Lasaga, have shown a curve derived from their model. It begins 100 million years ago in the Cretaceous, in the heyday of the dinosaurs. The concentration of atmospheric carbon dioxide was then seventeen times its present value. As you move toward the present, the curve falls steeply, to where, at 60 million years, it records a value very much like that of today. Then there is a gentle increase, so that at 40 million years the concentration of carbon dioxide stands at five times what it is now. Berner and Lasaga attribute this rise to a period of rapid seafloor spreading and to the accompanying volcanism and release of carbon dioxide. From 40 million years ago to the present, the trend for atmospheric carbon dioxide is downward again—the gas levels off at very nearly its contemporary value sometime around 20 million years ago. For 20 million years a balance has been struck between the breathing in and the breathing out of the Earth: The respiring of the volcanoes has been matched by the weathering of the crust and by the quiet crafting of shells.

At this distance the cone of Erebus was smooth and featureless. There was a delicate shade of pink, like tea roses, falling across its slopes. As I looked at it, I saw it differently from before. It stood now in the center of an imagined collage, like something I had seen in the gallery. Around it were objects scattered and indistinct: a crystal of calcite; a map of the world; a bamboo shoot; a night street under deep snow; a weathered crater in the Koolaus; a maple seed; a river bending by the red mill stacks; a cornfield in springtime; a wooden flume; a garden with a table; a molecule of carbon dioxide; a pink shell. Woven among these, in cycles, were the threads of the world, which were carbon and calcium, and these were overlaid by the cycles of yet a larger strand, which was water. And the water appeared to glisten, the way a droplet glistens on a needle of ice.

The Moat

We live at the bottom of a sea of air, and our sense of warm and cold are conditioned by the fact that we are surrounded by a mixture of gases. In gases, heat moves from one place to another slowly because the process requires that molecules travel for some distance through empty space until, by chance encounter, they collide and exchange energies, quietly and swiftly, as though they were exchanging confidences. This is why we can go into the winter air with just a shirt on, if only to collect another log for the fireplace.

In liquids it is not like this: The molecules, though free to move, are in constant jostling contact with their neighbors. Liquids have a different and more expeditious manner of transferring their energy, one that is more appropriate to the crowded spaces that define them. On the molecular level, it is a kind of hip-throwing, bumping action, a frenzied dance on a crowded floor, in which the excessive vibrational and rotational energies of one molecule are transferred directly to its neighbors and, in turn, communicated outward in all directions to the entire population. In this way heat (and sound as well) is conducted quickly and efficiently and over large distances, even though the individual molecules move hardly at all. Water, because of its remarkable tanglement of hydrogen bonds, is an excellent conductor of heat. The energy of a warm object in cold water drains away with the suddenness of a ruptured bag of seed.

▼ ▼ ▼

DURING OUR DAYS AT FRYXELL, the moat between the shore and permanent ice cover had widened to where we could cross it only by boat. When we set out for the lake, we roped the inflatable raft front and back with a long nylon line. I would move the boat off a little from the gravel shore, so that it just floated in the shallows. Mike would hand me an ice axe and a bamboo pole and a single paddle and I would be ready to cross. Perched on my knees, I would lower the bamboo into the water, lodge it against the sediments, and push. The motion was like that of punting, except that I was kneeling on the bottom of the raft, bent over to keep a low center of gravity, and the hood of my parka was up around my face. Usually the raft would drift down lake in the wind, and Mike would have to let out the line so that, at times, the rope would lie in a wide, somnolent arc on the water.

When I got into deep water and could no longer punt, I used the paddle, and when I neared the ice edge I drew out the axe. The ice axe was about three feet long, made of aluminum, and shaped like a pick, except that the helve came to a sharp point so that it could be used to probe for crevasses. We had taken it with us when we climbed the Canada, and it had helped us reveal in our path a dozen deep shafts that lay crusted over by only a few inches of deceptive snow. We could hear the dry chunks of ice that the axe had dislodged, falling away into the chasm beneath our feet.

Near the permanent ice, I would reach out over the prow with the axe and sink its pick end into the edge. Sometimes a huge chunk of ice that had been undercut and eroded by the lapping waters of the lake would break off and leave a fresh cove. There would be a tinkling sound as the ice collapsed and sank and then resurfaced. When the axe caught, I would pull myself to, so that the prow of the raft ran up over the ice, to where I could crawl safely onto it. Then Mike would pull the boat to shore and load it with samplers and bottles and supplies for the day, and I would drag it back out to the ice, unload it, and stack its cargo at a safe distance from the edge. We did this two or three times until we had moved everything onto the lake.

The coming of spring made life more difficult. The moat widened and the ice edge became weathered and treacherous. When we went onto the lake we knew we would be there for twelve hours or more, since there was no easy or convenient way off. It was frustrating to forget something, to leave a pipette or a vial of acid or the conductivity meter on shore. We had to check everything twice, so that we wouldn't have to trudge across the rugged surface of Fryxell and renegotiate the moat in our little craft.

One evening, when we were returning to shore, I noticed that the boat was not where we had left it. The wind had pushed it off the ice and it was floating over near the shore. The line lay coiled in the water only a few feet away from the permanent ice. I moved cautiously toward the freezing water, got down on my stomach, and worked my way to the edge, while Mike held onto my boots. As I reached for the rope, there was a sudden crack. I was in the lake headfirst before I knew what had happened.

The bright orange manual that the National Science Foundation sends you before you leave for the Ice speaks of survival in cold seawater. But there is really little difference between the frigid ice moat of Fryxell and the circumpolar seas of the continent. Both waters hover near the freezing point, though for seawater the freezing point is slightly lower. This fact, while possibly of great importance to the physical chemist, makes hardly a whit of difference to the dazed swimmer struggling to survive. "When the body is submerged," the manual says, "reflex contractions of smaller blood vessels briefly increase blood pressure and heart rate. The victim loses consciousness in five to seven minutes and will die in ten to thirty minutes. Body core temperature falls very rapidly."

I think I must have folded naturally into some strange position, because I was underwater, lying on my side. My parka was bunched over my head and my hands were reaching up, trying to grab anything, the ice edge, Mike's hand. I had this feeling that I was powerless to do anything. I felt my energies drain away.

In the blue, icy waters of Fryxell, an image of Sandy Beach appeared. It was Christmas Day in Hawaii and I was on my way home from an afternoon party at Keith's. I decided to stop for an early evening swim. The waves were rolling in at ten feet, rolling in one after another, flinging sand into the air with a mighty roar. I ran out to meet them. The water rose in steep walls that approached nearer and nearer until they blotted out the sky. I felt the water and the sand being sucked out from around my feet. For a second everything was water, curling and shimmering and plunging straight downward, roaring in over my ears, exploding, lifting me into the air. No sooner had I staggered to my feet from the first wave than a second one was upon me. I was gasping for breath when yet another wave hauled me into the air. I was breathing in the salt water, certain that I would not be able to drag myself even the few yards required to be out of harm's way. When I recovered yet again, I could sense another wave about to fall on me. I felt its presence more by the darkness that it cast than by its sound. With whatever strength I had left, I lunged forward. The wave broke on my heels and I felt myself being lifted in a great headlong rush by the spume. I was rolled up on the shore. I lay there coughing as the sea flowed around me and then drained back into itself.

The image of Sandy Beach shot before me in a flash and was gone. I began to struggle with my arms, to pull myself through the water. My glove touched something. It was the wall of the permanent ice. I was swallowing water as I moved toward it. I lunged for the ice edge, grabbed it with both arms, but it broke away into chunks and crystals. I lunged again but with the same result. I heard a voice. It was yelling, "The bamboo, the bamboo!" I saw something through eyes that were running like rivers, that were near blind with water. I took it and held it as though my life depended on it. I was pulled onto the permanent ice. All of this took seconds.

After I recovered from the shock, I took off my clothes and dried myself with a padded vest I had left lying in the sled. There was no wind. Absolute, merciful, calm. Mike gave me his checkered shirt and the thermal layer of his windpants. I lay in the sled on the permanent ice catching my breath. My chest was heaving, my eyes were wide open with adrenaline. I was remembering something Varner had said to me: If you were in real trouble, you would want Mike around. "I've only known him for a few weeks," Varner said, "but here's a guy you can count on. Believe me."

▼ ▼ ▼

Fryxell surprised us, but not again in unpleasant ways. From top to bottom, the water column was stratified. It appeared there had been no mixing events in the lake's recent history, no wind-induced turnovers or seismic palpitations, no turbulence of note since at least the day of Charlemagne's dream. We used the mercuric nitrate titration method to look at the chloride profile. There were the usual lovely color changes, from the blue-green of the indicator solution to the deep, evening-sky purple that the excess mercury made with diphenylcarbazone. Just before the endpoint, there would be billowing clouds of lavender drifting in aquamarine, then everything would turn, with one extra drop, suddenly. There was no mistaking when the titration was over.

We speculated that the evolution of Lake Fryxell occurred in this way: A great lake had once sprawled across the valley. Maybe it had stretched all the way from the ridge that fronts McMurdo Sound over into the narrow valley of Lake Hoare, and on up toward the Suess. Its waters, in the short summer, must have lapped high against the Kukri Hills and over against the hollowed boulders of the Asgards. There would have been an ice cover, though perhaps not as thick as now, and a wide moat in which the cone of Erebus would have lain reflected like the closed petals of a flower floated against the brown shoreline. Blue and brown and a droplet of white suspended in the moat and the unmoored summer ice moaning and shifting with the wind like the raft of a continent, drifting and seeking its place. Into the clear eye of the lake there would have been no one to gaze, no other living eyes, unless, of course, a small Adelie, formal and bent and treading the high shoreline, lost and groping its way toward McMurdo Sound, had peered Narcissus-like into the mirror of Fryxell and seen itself poised —its flippers extended over the water, over the clouds that passed below—as if it were on the verge of flight.

This large lake slowly evaporated, its water and ice went off into the desert air, into the green luminance of the glaciers and back to the sea. Calcite, gypsum, sepiolite—the usual sequence of insoluble deposits —rimmed the lake as their numbers were called, as the shrinking waters yielded them up as solids. In the end there was mostly sodium and chloride, dense brine, too concentrated even to freeze, and its number, too, was called. Abandoned by the waters, it dusted the frozen salt plains of Fryxell, lay with a white dry brilliance across the lake floor. It must have been a fierce patina under the night sky of Antarctica, with a slice of moon overhead or with the Southern Cross so near that it bathed the crystals with starlight. It is not possible to say how clear the air is here. How clear it was then.

When the summers are so cold that the streams no longer run, even for a day, the lakes disappear and turn to salt and dust. When it warms and the glaciers again sweat and pool in the sun, when the singing begins all over again in the mountains, and the plash of water can be heard over the wind the whole valley long, then the lakes return. Empty and full the vessel, like a chalice drained and replenished. Like a sacrament in the spoor of time. It must have happened that way at Fryxell. The water came in swift rushes down the deep channels, around the boulders that for centuries had tasted only wind and sand, and lay like a lens over the valley. And then more and more, year after year, until the basin had nearly filled. There were, of course, ions that came from the streams—sodium and bicarbonate mostly, but also calcium and magnesium and sulfate and iron and manganese and copper. There was diffusion too: the old salts in the lake bottom had deliquesced and redissolved, the ions had spread upward and outward like a gas released into the corner of a room. All of this you could see with a few measurements, a lowering here and a lowering there and a few quick titrations. The cycle of the lake through time.

But there was more. From nine meters to the bottom we measured increasing concentrations of sulfide. Near the sediments, the sulfide values were greater than those in the depths of the Black Sea. These numbers also told a tale: the sulfide in the deep basin had surely been produced by the rain of organic matter, by the carbon in the cells of algae reducing the stream-borne sulfate to sulfide. In the basin of Fryxell you could see the carbon cycle turning and turning in miniature, and with it, as a kind of Ptolemaic adjunct, the cycle of sulfur: carbon dioxide, organic carbon, carbon dioxide again. Sulfate, hydrogen sulfide, sulfate again. The inorganic and the organic, the nonliving and the living, intertwined, one rising out of the fevered remains of the other, then sinking back, then rising again. And these cycles would be linked with others. The sulfides with the metals—with copper and cadmium and zinc in the fine-grained settling of minerals; the carbon—the bicarbonates and carbonates—with calcium and magnesium. There was not a single destiny in the lake that was not tied to some other; not a single cycle revolved alone.

▾ ▾ ▾

On the day before Christmas the last helicopters left for the valleys at noon. Mike and I spent the morning packing up the camp and cleaning out the Kiwi Hut. There were no tents to strike, but as always there were the footlockers filled with samples, and the sleds and drills and meters. The plastic bags—"johnny bags," as they were called—containing the human wastes we had generated, all of the excrement and urine from our days at Fryxell, were placed into a separate box for transport back to McMurdo. We gathered bits of paper, a coffee cup lying in the shadow of a sandstone boulder, a half-buried spoon, a frayed length of rope, and put them into the pouches of our knapsacks. I swept the floor of the Kiwi Hut, turned the mattresses over to expose their fresher undersides, replaced the jars of vegemite and peanut butter on the shelves, and brushed the dried crumbs of bread and cheese and rice from the counter. Mike laid a chocolate mint on each of the beds, near where the pillows should have been. When we were finished, it almost appeared that we had not been there, at all. LEAVE NO FOOTPRINTS, a sign on the wall said.

I kept the radio set up until last, so that I could make coms with Hoare. I was sitting there in the sand, talking down into the valley and around the face of the Canada Glacier, where Varner and Walt were still tending the stream. The signal was as clear as I had ever heard it.

We had planned to meet at Vanda for Christmas, so it came as a surprise when Walt said, "We would like to stay at Hoare. How copy? Over."

"Copy loud and clear," I replied.

"Varner needs data, data, data!" Walt continued, speaking, as he usually did, for Varner, who wanted absolutely nothing to do with the radio once he had set it up and tuned it to working. Varner, who taught the electromagnetic theories of Maxwell and Hertz with the same fervor and sense of history that he accorded to Newtonian mechanics, shunned this tangible offshoot of nineteenth-century physics as though it were inhabited by ghosts. "Half the time it doesn't work," he said. "There are too many charged particles flying around down here. Too much atmospherics. Even Hertz couldn't deal with this."

I had wanted them to see the stark grandeur of Wright Valley and to enjoy themselves amidst the noisy holiday camaraderie of the little group that would gather at Vanda Station, but they were clearly taken by the tiny stream tracking the sun and temperature and the turbidity of the air, as it had for thousands of years, that was insinuating itself with its trills and motions and perhaps, by now, with its pattern of crescendo and diminuendo, into their favor. If this was so, I had no doubt they had made a fair trade.

Once everything had been put away, including, finally, the radio itself, there was little to do at Fryxell but wait among the heaped boxes and flight bags. I was near the circle of orange-painted stones in the center of which the helicopter ceremoniously puts down, and from that position I could see everything—volcano, glaciers, lake—that I had seen before, but now my eyes were focused on something else, something dark, a mound in the middle of the ice, a haystack of blue-green algae.

At one moment you see an object this way and at another moment you see it that way. It never really is the same object at all, except by the convention of naming. It is a different object with every glance, a different object in the associations that it calls to mind.

The cyanobacteria, which I could see only as a dark, pustulent stain on the lake, were something more than just an awkward physical impediment around which I moved my sled. They were the visible end of one process and the visible beginning of another. I knew little about their physiology, but I knew their origin and I knew their destiny. The cyanobacteria were photosynthesizing organisms, much like those that had colonized the ancient, anaerobic Earth. Their forebears had transformed the planet's atmosphere, some three billion years ago, from a dense and, to us, unbreathable canopy of carbon dioxide and water vapor, to one leavened with oxygen. Those organisms, as George Simmons had shown, colonized the sediments of Lake Fryxell. They grew in lush mats on the bottom, and to the divers swimming through the cold blue waters of the lake, they looked like fields of grain. They took inorganic carbon from the water, and with the energy of a six-month sun, turned it into carbohydrate and protein and fat, into DNA and RNA, into reduced and energy-rich molecules.

There was a kind of exotic flotation process at work in the deep waters of Fryxell. When the bacterial mats made too much oxygen, or made it too fast, so that it collected in gaseous pods beneath them, the gas would buoy them up from the sediments and send them rising in huge clumps through the water. Their upward journey was swift and brief—a kind of balloon ascent through the sky-colored waters of the lake—and was arrested by the base of the ice sheet. The mats would lie there, pressed against the undersurface. Then, as a new layer of ice froze to the bottom—to replace the layer that had been whisked away by the wind on top—they would be trapped. Slowly they would move upward through their crystal sarcophagus, until finally, after twelve years more or less, they would emerge into the sunlight of the lake's surface. Desiccation followed, and then dispersion by the winter winds across the lake and across the valley. As though they were autumn leaves, the stromatolites began to disappear into thin air, began to turn, in the slow distilling processes of decay and rot, back to carbon dioxide and water. The cycle, through all the molecular fixing and arranging and building, all the linking and breaking of bonds, the silent electronic workings of chlorophyll-a, the shifting and ionizing of water, was completed. The dark mounds that lay on the ice of Fryxell drifted back to the air from which they had come, a reenactment in miniature of the whole planet's respiring.

It is not much different from the leaves that lay scattered everywhere when I left home, little bowls of mildewed brown amid the winter-green grass. In October they were still orange and red and magenta even as they lay open and veined on the ground. But as the weeks went by they curled and blackened and shrank in on themselves. Their season was over. They were going back.

All over the world it is like this, this breathing in and breathing out. Before I took down the antenna and folded the radio into its canvas bag, I heard voices coming from the South Pole. They were clear and they were talking about carbon dioxide, about the latest point on the Keeling Curve. Not that there was anything extraordinary about this one point—it was where it should have been, given everything that had come before. It was what had come before that was extraordinary.

All over the world, this shallow breath of the Earth is signed upon the air. The signature rises and falls through time, in a sinuous curve. You could not mistake it for anything else but the Earth's breathing. It is the curve of carbon dioxide through modern time, its concentration in the atmosphere, plotted month after month—the Keeling Curve. From the slopes of Mauna Loa on the Big Island to the desolate wyomings of Pole Station—where the sleepless recorders work day and night inhaling the air, taking its measure—the patterns are the same: In the spring of the northern hemisphere, the carbon dioxide of the atmosphere declines as the trees unfold, as they imbibe water into their roots and carbon into their leaves, as the warming seas ripen with phytoplankton off Peru and California and Cape Cod. The sweet yearly inhalation and budding, fecundity and opulence and color everywhere spreading like a blush and holding. All through spring and summer the Keeling Curve falls. In autumn and winter it rises, when in fields and empty lots, in forests and swamps, and in the gray winter of the sea, carbon dioxide is returned to the air.

But the curve is not a simple wave, evenly etched and proportional, as one might have expected if the Earth were merely returning, atom for atom, its springtime carbon into the fallow ruck of winter. Instead it moves upon a rising line, each winter's carbon dioxide higher than the preceding spring, and higher than the winter before. It has been like this since the first days of the measurements, since the first crude chromatograms were taken by Keeling on the slopes of Mauna Loa volcano more than three decades ago. It has probably been like this, this slow rising, for two hundred years. The ascent of the curve, here and everywhere else on the planet, is our doing. It is traceable to distant lands and ingenious craft: to the making of iron from coal (learn to transfer electrons from carbon to iron and everything becomes different, as momentous as splitting the atom); to the recovery of oil from its oozing pools in the dark earth; to the clever means of movement we have devised. Who would have thought that these things, these techniques, could change so much?

When you run your finger over the Keeling Curve, you feel slightly off balance. Not the vertigo that attends a whiteout, nothing that dramatic. But just the troubling sensation that something is not right. It is the kind of sensation that you get from seeing a signature that has changed a little, that is no longer quite straight across the page, no longer as steady or as flowing as it used to be. It is like that.

Sometimes you can wait for the helicopters for hours. And sometimes the wind in the mountain passes can deceive you and sound so much like the rotor blades that you stand up and look around. But it is not as bad as it was years ago when there were no field radios, and you could wait for days, thinking, *They'll be here any minute now. I think I hear them over the ridge.*

I did not object to these delays. They were little windows, little rooms, from which you could take notice of things. I had heard the people from Pole Station talking on the radio about methane, common "marsh gas," the stuff of will-o'-the-wisps, the simplest compound of carbon and hydrogen. They were saying how their instruments were recording a gradual rise in its atmospheric concentrations at the Pole. The rise of methane was being observed elsewhere too, in Hawaii and Samoa and at other remote stations around the globe.

Looking over the wilderness of Lake Fryxell and thinking about what I had heard—the reports on carbon dioxide and methane, a comment from one of the scientists at Pole Station on the latest ozone values, all of which had an immediacy here that they might not have possessed elsewhere—I thought of how, just a few hundred years ago, in the grimy factories of Coolbrookdale, in Dalton and Rutherford's own Manchester—places where we had collectively learned to reduce the rough oxide of iron into lustrous iron metal through the agency of carbon; where we had learned to fashion metals into bold engines of change; had devised beyond cunning, in the rooms of great genius, ways to turn liquid water into steam; and with throttles and cams and valves and flywheels, learned to control it, and then to use it to usher forth carbon in ever greater abundance from the Earth, raw mountains of anthracite heaped and gleaming in the moonlight, the way I remember it in Pittsburgh, passing on barges through the channels of the wide rivers, under the girdered bridges flung in perfect engineering arcs up and down the Indian-named waters—Ohio, Allegheny, Monongahela—passing toward the mill stacks reddened with volcanic fire, toward the ladles and ingots and cranes of the mills, toward the rising clouds of carbon and white steam and sulfur. Two hundred years ago we were setting off from familiar shores, where the fields rolled in verdure toward the towns, where the days were marked in prayer—in Matins, Lauds, Prime, Vespers, Compline, where death came early and in mystery, unexplained, an act of God, where the wooden ships lay crushed in ice. Two hundred years ago no one had yet tapped the source rocks, the shales and limestones, for the lakes of oil that lay within them—all of that energy taut in the covalent bonds of carbon and carbon, of carbon and hydrogen, awaiting escape as in one huge conflagration, awaiting the chance to turn a wheel, as though the huge ages past—Carboniferous, Permian, Cretaceous—were nothing more than the kindling of our own.

It's as if, two hundred years ago, we pushed off from the known shore. Our instruments were crude. We had no idea where we were going, how wide the moat of this strange passage might be. We had no idea this crossing would change the world. Two hundred years, more or less. Here in the lower Taylor Valley the change is nothing. Here you can count the boulders that have moved in that time, you can reckon how far they have gone, how much the lake has risen, how deep the streams have weathered into their channels, how much the wind has cut the face of the stone, how much the glaciers have inched backward, melting. But you need fine instruments to do this, the changes are so little.

At first I thought it was a skua. I naturally associated any motion that was not ours with the large gulls that came infrequently into the valleys in search of our food. But it was the Addie. I watched her through my field glasses. She was walking along the north shore of Lake Fryxell. There was no point on the whole circumference of the lake where she could cross over to the permanent ice. Everywhere the moat was too wide. She was moving quickly, moving down toward the Commonwealth Glacier. I knew that when she got to the streams she would find places to cross. Then she could turn south and east and walk up the moraines and out to the Ross Sea.

I traced the path she would take with my field glasses, traced it all the way over to where Erebus was steaming into an overcast sky. I picked up a small orange dot on the horizon. It was the helo. We had been waiting for it for three hours.

▼

EIGHTEEN

Merry Christmas

My last memory of Vanda Station was the photograph that Canfield had taken some years ago. It was taken inside, in black and white. There were seven of us seated around a long table whose surface was dark and smooth. There were large glass ashtrays and empty cans of Steinlager. There were dark bottles of Tia Maria and Drambuie, a tapered vase with a small bouquet of white artificial roses, and two cameras. In the photograph no one is asleep, but there is no conversation. No two faces are turned toward each other. Matsumoto rests against the back of his chair, his hands folded on his lap. There is a diver's watch on his wrist. His two Japanese colleagues lean forward a bit, their elbows on the table. There is an Australian scientist seated in the corner, his head tilted up toward a shelf of geological reference books. A large photograph of a dark-haired woman, her back toward us, hangs on the far wall. The photograph is larger than the window. You can see the woman's delicate face in profile, her chin tilted slightly downward, reflected in a mirror. She has a diaphanous shawl draped around her shoulders, which forms a small V at her neck. In Canfield's picture the window is opaque. You cannot see out. But the lake is just beyond it, and so are the peaks of the mountains.

▼ ▼ ▼

VANDA STATION IS ON HIGH, rugged ground overlooking the lake. The helicopter lands in a small bowl of stones at a level that is only slightly elevated above the present shoreline. A narrow path, mostly rutted sand and pebbles, runs a tenth of a mile up to the station from the landing site. Halfway up the rise, the path skirts a plywood outhouse that is open in the front to the wind and weather, and is so situated that its visitors are afforded one of the most spectacular views of a mountain to be found anywhere on earth.

A little farther along there are barrels of fuel oil and discarded drums of waste, which, at the season's end, will be flown back to McMurdo. As you approach the green structure that is the station itself, you see the sun-bleached bones of a crab-eater seal lying atop a mound of rocks. Beyond this are a few smaller buildings—a carpentry shed and two sleeping quarters. Wind generators are perched high on their towers—one near the station, the other down by the lake. There is an orange sign over the door that reads VANDA OASIS. From the front of the station you look east toward the alb-white glaciers dropping toward the sinuous channel of the Onyx, west over the gentle furrows of the lake to the ice-chiseled Dais, south to the russet sandstone steeples of the Asgards, and north onto the broad white flanks of the Olympus Glacier. Beyond the tiny scope of the station, there is nothing of human making to be seen. It has always been like this.

Inside the door is a small vestibule crammed with shelves of food and supplies, much like the Kiwi Hut. Three people standing in this space, all removing their parkas and bunny boots and attempting to get settled at the same time, will seem like a large crowd. From the vestibule you go up one step and through a freezer door into the main cabin. It too, by any conventional standards, is a diminutive and roughly appointed living space. But after the cramped tents and huts of the Taylor Valley it seems a generous and substantial room.

Though there had been a few changes, things were much as I remembered them: the wide sink and the cutting board facing the window onto the Onyx; the bookshelves; the long table; the ample light on the wooden floor. There were Christmas decorations everywhere. Strings of silver tinsel, bedecked with Christmas cards that had come from schoolchildren all over New Zealand, ran on diagonals the length of the room. In the corner was a tree, a freshly cut pine from the alpine mountains of the South Island. A young Japanese woman who was working with Dr. Torii had hung multicolored origami cubes and intricate bluebirds and starfish from the branches. Quavering strings of metallic tinsel floated against the tree in shallow curves and draperies like morning webs slung with dew. A golden-haired angel traced a small curve through the air against the backdrop of Vanda.

Twelve of us had gathered at the table for dinner: Moth, A.Q., and Lou—the Kiwis, at whose table we were guests—Drs. Torii and Kyoku, Mike, Tim, Dr. Yu, Alan Campbell, and I, and two New Zealand geologists who had been off in the mountains for two months studying the motions of glaciers. Dr. Torii was pouring sake into liqueur glasses while A.Q. carved the turkey he had just taken from the oven. Lou, who was the leader of Vanda Station for the season, was wearing a starched white shirt and a black bow tie he had brought with him from Auckland in September especially for this occasion. Moth and A.Q. were also dressed in white shirts and ties. Those of us who were visiting made do with whatever appropriate bits of material we could find. Dr. Yu, whose blue shirt was buttoned at the neck, wore a bright green ribbon beneath his chin. Tim fashioned thin sheets of lead into bow ties for himself and for Mike and me. We attached them to our shirts with paper clips. Kyoku had changed into a colorful silk dress that swept around her ankles when she moved.

Dinner came steaming and all at once. Porcelain plates were heaped with green peas and baked potatoes and long, moist strips of white turkey. There were dark red wines that swirled with the color of garnet against the rims of leaded crystal, and a pale Chablis that was tinged with amber. The mingled aromas of food, of cranberry and asparagus and gravy, were so thick and rich in the air that they seemed on the verge of condensing on the cold windows of the station. Candlelight danced in the silver concavities of the serving trays, and the sounds of brass, of trumpets and horns, lifted in praise from a speaker in the corner of the room: "O come, O come, Emmanuel. Redeem thy captive Israel."

Despite the festive atmosphere, the conversation was subdued— laughter and long pauses, the clatter of utensils breaking the silence. It was the awkward conversation of strangers, come together out of empty mountains and valleys and snowfields, glad for one another's warmth and smell and scrubbed faces, but not wholly there. Distant. "So, Mike," one of the Kiwis would say, "how's the old hut over at Fryxell holding up? Never been there myself." And Mike would say a few words and then throw in a wry comment on how the vegemite could have been better. And everyone would laugh. Or there would be toasts: to the Yanks, to the Kiwis, to the Japanese, to Dr. Yu and all of China. And everyone would stand and shout "Gambay," and then there would be silence again and the clank of metal. And then another question and laughter and silence.

The smell of the pine came and went as though it were a river winding invisibly through the air. When it crested near me, I would be carried away by it. I would be seated in a room and would hear my daughters' voices approaching, the shushing of their full-length pajamas against the carpet, and then their beaming faces. They would want a story before bedtime, by the tree with its soft lights. And I would invent something for them out of thin air—about a penguin I had known, about a strange land I had once visited. And Wanda would play her guitar for them and sing to them in languages they had never heard before.

Or the river would crest near me and I would be carried back further. I would remember the way the altar looked with its freshly cut pines, as I stood there, a small boy, next to my mother, who held a long taper in her hand, held its fire poised over a tall blue lamp. When it touched and lit, I could smell the sweet carbon wax rising into the church air. And we would pray Hail Marys and bless ourselves, rise and walk to the huge doors, push them out against the drifting snow that had gathered on the stoop. The choir would be practicing for the next mass, and we could hear them as the door closed slowly over the sound of their voices: "Veni, Veni, Emmanuel." Then we would walk home in the middle of the streets, for the snow in those winters was too deep for cars.

For dessert at Vanda there was plum pudding flamed with brandy. A.Q. brought it to the table on fire, amid loud applause. When it came, I thought of Thomson and his electrons, of his improbable plum-pudding atoms bristling with charge. Dr. Torii poured sake into tiny glasses and we toasted again.

After dinner there was a nodding warmth and heaviness that fell over the station, and people were slowly sinking like soft sculpture against the walls. Moth called it "high gravity." "We're in a high gravity zone, mates," he said from the corner, smiling, almost asleep himself. Mike, who appeared on the verge of sleep in his chair at the long table, his head bobbing above the remains of the plum pudding, suddenly looked up. He made a tired overhand motion with his forearm and wrist, his fingers spread far apart, as though he were making some kind of portentous last request. We had brought a good leather football with us from Ohio, and I knew he was suggesting that we get some exercise, a game of catch, before we succumbed to the remarkable forces of indolence that were pulling us down. "Ah, great idea," I muttered out of my own stupor, and we made our way slowly to the vestibule to get our parkas.

Outside, there was a stiff wind blowing in from the west, from the plateau. Near the door of Vanda Station you could see down the length of the lake to where the Dais was caught in a cloud of fresh snow, to where it looked grainy and stark, like Half Dome in the black and white photographs of Ansel Adams. We were throwing along the pavement of boulders and flat stones that covers the moraine, that rolls in lifeless hills and valleys to the scree slopes of the Asgards. When the ball came out of the east it traced a sharp curve through the air as though it were under the influence of a great geostrophic force. Sometimes it would appear to stop altogether and just hang before me for an instant; then it would float backward and begin to fall to the Earth. "Coriolis," I joked, and threw the ball downwind toward the Onyx. "Magnetic field," Mike bantered back, "it's the big magnetic field." It was the farthest I had ever thrown a football.

We had been running and throwing for twenty minutes, and we were wide awake. I heard a sound coming from the direction of the river, but it was not water. When I looked, I saw a Huey. It was coming in low and fast up the valley on the far side of the river.

The pilot had brought mail. Mostly it was for the Kiwis. But there was a letter addressed S-041. It was from Canfield. My hands shook a little as I opened the letter. I was seated at the long table with Dr. Yu. Tim and Mike were leaning in, peering over my shoulder. The pilot and his crew were standing by the oven warming their hands, waiting for the batch of fresh date scones that were about to appear. The others had gone out to the bunkhouse. I could smell the pine over in the corner, and hear the wind in the guy-wires outside. A light snow had fallen on the umber basalts of the mountains. The letter began, "Merry Christmas."

The nature of rivers is to take and to take in, to dissolve into the flow and carry away. Thus the Huang He, in the heart of China, transports a billion and a half tons of windblown silt in a single year until, in its lower reaches, it moves with the torpor of molasses. The Colorado cuts a mile of stone straight down, as though it were a knife on wet clay, a whole canyon in a mere million years; and a small stream in the Wasatch Range once hefted a boulder of ninety tons for a mile along its flooded course. Water falls down the face of the land and carries the land away, etches and pits it and sweeps it in great suspensions and solutions to the sea. By three inches in a thousand years, on average, are the continents lowered by this power.

The Onyx, except for its contrary movement from the sea and its short and erratic season of flow, was not much different from ordinary rivers. It mined the scratch of arid earth that was its domain, the way all rivers do. Out of the rock, without instrument or plan, it plucked the metals and carried them to the lake. Everything was carried to the lake.

We have mined and worked these metals for thousands of years: copper in the Middle East, ten thousand years ago; its extraction from ore in Persia, seven thousand years ago; its casting, with tin, as bronze in the valley of the Huang He, three thousand years ago. Iron, 3,500 years ago; its conversion, with bits of carbon, into steel, one thousand years ago. And while metals are the hallmark of civilization, the stuff of its weaponry and ritual and fine art, the material of its coinage and commerce, the namesake of its Ages; and while they are hard won with the sword of fire from cinnabar and bauxite, rutile and galena, while the last centuries have seen the dark mines of nearly every land hung with lamplight and sweat; while all of this energy and industry, genius and greed, has poured into the shafts and mountains and foundries of the Earth and liberated metal from it, the streams and rivers have always gathered into their waters schools of copper and zinc, flights of cobalt. This has been so for billions of years.

It was true of the Onyx. It was all swept into the lake, from which there was no exit. In a mere thousand years the Onyx had delivered tons of copper alone. The coinage of tribute. Yet it was not in the lake. Where had it gone?

Science is all conjecture and refutation, as Karl Popper said. There is mystery all about. Things unexplained. Burning questions that sometimes burn, perhaps inexplicably, for one alone. I wanted to know where the metals had gone, right here, in this one lake, not much larger than Walden, yet intensely and austerely more beautiful; this one lake, in all the world, completely removed from art and commerce and industry; its shores unsettled and unvisited; its howling emptiness in this valley, for which "wilderness" is far too tame a word. When I told her these things, my mother, who found a Latin phrase for everything, said, "Vanda is *sui generis;* it is its own kind." So were all the lakes.

We had our conjectures. Possibly the algae were sequestering metals, taking them in as intimate centerpieces in the molecules of life: iron, for example, the way it sits among the nitrogens of the heme group, a kind of anchor point in the mosaic of five-membered rings that lie in a plane around it. This iron is the active center of heme, itself the center of cytochrome, which is the electron carrier involved in photosynthesis. Manganese and cobalt and other so-called trace

elements all play crucial roles in the activation of enzymes, the molecules of plant and animal cells that serve as the catalysts of biochemi-
cal change. Copper is involved in oxygen transport, molybdenum in
the capture of molecular nitrogen from the air. Carboxypeptidase A,
another enzyme, in which zinc occupies the central active site, is
designed to sever only the terminal amide bond of a peptide chain.
Such specialization! The cutter who cuts the bolt of cloth at the last
blue flower only.

Thus the cycles of a lake—the swift cleavage of carbon dioxide
from water, the remarkable four-electron reduction of oxygen by
which respiration occurs, indeed, the great geochemical cycles of the
world—all turn by the grace of these trace metals, these nimble accep-
tors and donors of electrons on whose partly empty d-orbitals so much
else depends. Mere bits and pieces of matter, mined so assiduously
from stone, stolen from rock and taken into the chambers of life for a
short time and then released again. Without the trace metals to speed
us along, perhaps nothing much would happen. Things would not
grind to a halt so much as just wither. The weeds in the sandlot would
slowly, imperceptibly, decay to give rise to . . . nothing. The algal
bloom, the soft green blush of the lake, would die and settle, never to
be returned to the light.

And yet when their concentrations are too high, these very same
metals, upon whom every life process depends, wreak havoc on all they
encounter. The poisonings and cruel deformities that we associate with
the name Minimata Bay; the "ouch, ouch" disease of the Jintsu River
basin; the anemia and mental retardation that afflict cities; the buildup
of lethal white crusts on the gills of fish; "mad hatter's" disease; the
death of aquatic plant communities; the contamination of whole estu-
aries and inland seas—all of these can be caused by an overabundance
of trace metal. So the balance must be exactly right, the cycles must
turn at their preordained speeds, not too slowly, not too fast.

Metals have always coursed through the veins of the Earth, liven-
ing its duller tones of brown and gray with splendid hues: the soils of
Alabama, which burn like cool fire beneath the green pines and blue
sky; the red beds of Australia; the clays of Gubbia; the rusts and roses
of the desert—all of these are imbued with the subtleties of iron. The
purple of amethyst quartz is due to manganese and the green of mala-
chite to copper. Ruby and emerald contain only traces of chromium,
but this atom, with its d-orbitals, its laddered rungs of energy, is

enough to turn pallid silicates into pyramids and prisms of magnificent light. These purely inorganic washes are as variegated and beautiful as the colors of life itself.

But in very recent time—through the enormous powers bestowed on us by carbon and steam and by all the inventions and processes by which these are put at our command—we have so quickly taken metals out of the Earth and dispersed them willy-nilly over land, sea, and air, that it might seem we have been engaged in a kind of mineralogical diaspora. What was once locked in the wintry hills is now diffused by the waters of a thousand streams, by the currents of a thousand winds. Who would not wonder where these have gone?

There is a kind of conspiracy at work, Karl Turekian said. It is worldwide and, as far as we know, benevolent. In every lake and ocean, in every parcel of atmosphere, there is a cleansing that tempers the Earth, that drags it back from squalor, that countervails its self-undoing. Metals pour into the lake, but the lake removes them. Metals pour into the sea, but the sea is not full. The volcanoes breathe sulfur and chlorine, but the rains and winds carry them to Earth transformed. "Nature is a constant worker," said Newton of his newly limned world of matter and motion and force. The geochemist and geologist cannot but agree. But how quickly? How quickly, and by what mechanism, does the cleansing work? Is it fast enough? Is it being overwhelmed?

I read them Canfield's letter. It was about cobalt and lead, about the "conspiracy" that had led to the fates of just these two.

There was a pie chart for cobalt. It told how the metal was partitioned throughout the particles we had just sent him, the particles from the forty-five-meter trap. On the chart there were five sectors representing each of the possible phases into which the cobalt might be partitioned: there was a sector for the cobalt that was bound with organic matter, with the dead husks of sinking algae and bacteria; there was a sector for the cobalt bound into crystals of hematite and amorphous hydroxides, substituted there for atoms of iron; there was a sector for the cobalt that had been precipitated with calcium carbonate, a visitor, an interloper in the calcite lattice; and there was a sector for cobalt adsorbed onto surfaces, stuck there like a burr on a woolen coat. And then there was the fifth sector, the sector that symbolized cobalt that was tied to manganese, that was being removed from the lake by the manganese oxides.

My eyes blurred when I first looked at it. So much information compressed. I was thinking of Keith on the lanai as he contemplated the ocean, as he held his glass to it in a broad salute. How he had said the proof was in the particles, in the hard-won chemistry of the particles.

Tim saw it before I did. He saw the wedges of the pie lying there before us. He saw the angles the sectors made around the central point, the way they opened out, how some were mere slivers, and how others spread wide like a proclamation. "It's the manganese oxides," Tim said. He almost whispered it; his voice rose on the last syllables as though it were a question he was asking, as though an assertion would just make the figure go away.

"Wonderful," I said. "Wonderful," as my eyes finally focused, as the geometry of the figure slowly began to transform itself into chemistry, began to reveal the lake, the joined lives of the elements.

"It all makes sense," Tim said. "The way the cobalt seems to track the manganese in Vanda, the way it increases when the manganese begins to dissolve."

"And the Pacific nodules," Mike said. "The way they're so rich in cobalt, almost ores, really. It's remarkable." He was running his finger around the pie the computer had drawn. "The data are all saying the same thing. The lake, the sea, our particles—it's all one story."

It was only one data set. A single metal. A single trap. A single lake. But it was Christmas night, and Canfield's data was always good. I wanted to believe.

Mike was over by the sink now. He was pouring wine, pulling small dishes from the shelf. You could hear them clatter, blend in with the creak of the windmill outside, with the deep wind moaning in the valley. He had a knife in his hand. You could smell pumpkin and cherry and mincemeat in the air. He said he was taking orders.

Among geochemists there seemed to be a curious and widespread belief that we really understood nothing at all about metals unless we understood lead. In consequence, it acquired a kind of celebrity that few other chemical elements could ever hope to claim. Possibly this was the result of a spillover from the popular press, where lead often stole headlines from corporate takeovers and from successful invasions of small countries. Copper, nickel, zinc, silver: beyond the commodities reports, these were scarcely mentioned by anyone. No one cared; Bunker Hunt, maybe. But lead dominated local news, commanded documentaries, was even the subject of famous novels. Geochemists circled helplessly in its fame.

There is more to lead than its tabloid cachet might suggest, however. Lead is civilization's trace metal of choice, and it is everywhere. There are estimates that all natural atmospheric sources of lead, that is, lead from windborne soil particles, from bubbles bursting at the surface of the sea, from volcanoes and forest fires and assorted biological processes, contribute twelve thousand metric tons to the entire global cycle of the element. It has probably always been this way, give or take a few thousand metric tons. Unlike copper, however, this number is insignificant in comparison to the burden of human commerce—lead from mining and burning and smelting and manufacturing—that is added to the atmosphere at the fierce and gluttonous rate of 332 metric tons per year, nearly thirty times the natural flux. Of all of the trace metals that yearly rain or settle from the skies onto the surfaces of lakes and rivers and oceans, lead is by far the most abundant: 100,000 metric tons per year, in atmospheric fallout alone.

Lead is everywhere. Heaps of it have been vaporized, sent whistling up the stacks of Renaissance towns, Enlightenment cities, Victorian mills, and now through the leprous pipes of freeways, across fields of barley, wheat, rye, and corn, into the root and fiber of living plants. Like the Romans with their tainted wines, we eat lead, store it happily in our bones. It is the stuff of dark mines, unearthed and scattered on the winds and tides, come secretly home to us. The highest concentrations of atmospheric lead are, of course, near the industrial cities of Europe and eastern North America. But lead hovers, too, in the sprawling haze above the Arctic and mingles with the distant surface waters of the Pacific. The lead content of coral shells has increased fifteenfold in the last hundred years. Even the lead content of Antarctic snows is four times what it was in the centuries before the Industrial Revolution.

I told them what Canfield had found in Vanda, in the water samples we had sent back from McMurdo. How he had had to concentrate the samples twentyfold just to measure the lead, the concentrations were that low. But he had found patterns, and in them you could see the way the lake worked, how it cleansed itself so fastidiously of even the small traces of lead that the Onyx brought to it. You could see the natural cycle of lead, and you could see it, for the first time, unobscured. The profiles for lead were clear and unambiguous and they were linked to manganese, just as cobalt had been.

▼ ▼ ▼

Beneath the ice, in the first blue water that the Kemmerer touched and enfolded, there was only a fraction of a microgram of lead per liter. And below there was even less. Ten meters. Twenty meters. Thirty meters. Forty meters. Slightly less each time. You could imagine the sinking oxides of manganese gathering lead, harvesting it onto surfaces, storing the atoms one by one, walling them off, cementing them in fields of charge, carrying them down and down to where the polar light grew dim. But then something happened. At fifty-five meters, just where the manganese oxides began to crumble and dissolve, to fade into ions, to enter as charge into the liquid state—just where the manganese oxides seemed touched by madness in the changing conditions of the lake—the lead was released. At fifty-five meters Canfield had found clouds of it. At just this depth, where the manganese oxides could no longer hold it.

"But that's just where the lake becomes more acidic," Tim said, looking at one of the graphs, sipping his wine. "That's just where the pH begins to fall off. In slightly acidic waters, the lead is set free. It must be that way everywhere, in every lake."

There was more. At fifty-eight meters the cloud of lead ions began to thin again. Something was removing them from the deep lake. "It's the organisms," Mike said. "It's the algae and bacteria. It's warm down there, as warm as it is in this room. Remember Armitage. How he had felt the warmth on his fingers. Things can grow there. And the nitrogen and phosphorus are plentiful, they're diffusing up, bringing nourishment all the time. And of course there's still light. The water is so clear."

That was what Canfield had speculated, too. It must be the organisms at that depth, taking in, capturing, sinking, removing. And deeper still, there was even less lead. It had all but vanished from the samples. "Sulfides," Mike said, wrapping his large hands around his cup. "Lead sulfides. The lead has been precipitated, turned back to ore. To galena, whatever."

In Canfield's data there was a story, an image, a sensation. Lead in a distant lake, in a distant land. Lead dissolved, transported, captured, released, transformed, captured again, turned back to stone. Alchemy. Here, I imagined, you could see it, vividly and clearly.

You could see the Earth in miniature, this veil of purity sweeping across its burning waters, the lake like a dazzling shield against the light, against the dark mountains, against the saintly winter combed in snow. You could see the veil, secular and falling in the evening air, in the sea, dragged and combed and caught out in the wind-made spume, sweeping the cloister's marble floors, the swirling dust silent and rising. Down the center of the valley, its very axis the lake, you could feel it as breeze and ice and eternal dawn, you could feel yourself living forever on its taste, on the very sound that it made over the stone. Sometimes, though, it moved as columns upward, the way the air rises from a lake's center, the way the church light seems to move heavenward, lifting along the colonnades into the arches, joining stone to stone, shell to shell. And then you would pause and sit and be alone on the lake, on its smooth ice, on its surface on the floor of the world, the ribbed mountains with their ancient time, their passes and boulders and memories, and you could feel it all working like a breath and a breathing, like the draw of the spirit down and over and up and over and down yet again, a gyre, a cycle that sped and whispered and cleansed, that turned it all pure again. "The wind goeth toward the south and turneth about unto the north." And with the wind and with the water whirleth that which cleanses and renews, that which redeems and makes whole.

It was not long before the station was nearly empty. The pilot and his crew had flown back to McMurdo, and most of the others had gone to bed. Dr. Yu was at the table playing a solitary game of chess, and A.Q. was strumming his guitar. I was sitting there with my journal, enjoying one of those rare moments when time had stopped.

A.Q. was singing a slow Australian ballad, odd for a Christmas night, about Gallipoli. It told of a battle in a recent war. A war that began two years after Scott's death. How the troops came, wave after wave, singing, out of their holes. How they knew their fates even as they rose to their knees. "Waltzing Matilda, waltzing Matilda, you'll come a-waltzing, Matilda, with me." A.Q.'s voice filled the little station with the refrain. Dr. Yu had turned toward the window. He was looking onto the mountains of the Wright Valley, the mountains of Antarctica. You could hear the blade of the windmill creaking outside in the wind. It sounded almost like accompaniment. Dr. Yu seemed to be in another world. I didn't know whether he understood the words or not.

Alan Campbell was working at his easel. He was sketching something he had begun up in Bull Pass. In the painting, which was filled with the color of the stone and the color of the river, I could see time laid out before me in turquoise and tints of plum: the time of the river, a few thousand years; the time of the lake, ten thousand years; the time of the mountain, a hundred million years; and our time, the time of tents and instruments, of small measurements recorded in books, of blue flames and strong coffee, the smell of pie, the time of science and art. Not much.

By the window at the end of the hut, where the photograph of the beautiful woman had once hung, there were news clippings now. They were from the Auckland paper. They talked about Trevor Chinn's work on the Onyx, how the flow of the river had been increasing year to year, since he began measuring it, how it might all be tied to carbon dioxide. "An early sign?" the paper conjectured. "An early warning?"

NINETEEN

Cathedrals

We are asking how you make possible a sense of the sacred, how you stir up a feeling of awe, how the nudge of the transcendent can be sensibly perceived. To be edified means to be drawn out of oneself and made to feel part of a priestly people, a holy nation, offering an eternally valid sacrifice and receiving life-giving gifts. But all of those perceptions are mediated through the senses; they are the joyful children of sights and sounds and smells. Tied to liturgical content and enmeshed in the world of sense, the hallmarks of edifying liturgy are tranquillity, prayerfulness and conversion.

Eugene Green

▼ ▼ ▼

THERE WERE NO COLORS HERE like those of the night. The mountains were plum with iron-rich stone. The glaciers had passed into liquid silver, the flat paving stones, gathered in pathways and broken roads along the valley floor, were golden. Red basalts floated from high above in the curved moat, drifted to the edge of the permanent ice. There was no wind; you could hear the river gurgling among the boulders, sweeping the clear sand. If you closed your eyes, you could see the ions of Arrhenius move in the cold current toward Vanda, you could see the flickering clusters of Henry Frank form and dissolve, split as they streamed past the stones, re-form and break again. Blue *is* light," Henry Adams had said of the blues of the Chartres cathedral. The sky was like that. The moat, which was a mirror of the sky, ringed the lake like a pane of pure glass. Except for the Onyx, it was still. The whole assemblage of the mountains rose with parabolic grace from the river's course. Towers of stone, wind-carved and ornate, cast their treeless shadows on the slopes.

I knelt by the river and took a stone from the deep pockets of my windpants. It was a black ventifact *("ventus facto,"* she had said, "wind made"), an igneous rock, once extruded on fire from the deep Earth. During eons of exposure to the wind, its surface had become smooth and polished. It was black, it returned no light. Several years ago I had brought this rock home, thinking it would remind me of my days in this valley. But it soon became covered with dust on the shelves. I rarely noticed it, and even when I did, it seemed out of place. So I carried it here. I found the spot near the river where I had collected it. I laid its rough base, covered by a patina of salt, in the current. I faced its smoothest side toward where the winds would soon come, toward where they had come for a million years. No sooner had I placed it than I saw a slight milkiness run out from beneath it, a thin wisp of carbonate moving into the flood. On the banks of the Onyx, the season's work nearly done, I felt as though I were in the silence and light of a cathedral. I felt very small, in the presence of something whose name I could not speak.

Our town's cathedral, as I remember it, was no more than a pale copy of Chartres. It was the best that a town of Irish and German and Eastern European immigrants, come to make steel and a life, could do. But even in its blackened state, with all the airborne wastes of the mills, a generation of sulfates and acids, of the carbon soots laid black upon the white stone, it was beautiful. As you drove in from the city and looked down the wide valley that the small stream had carved, the towers rose over the town, graced it with their magnificence. In ornamentation they were simple, more like the octagonal north tower of Chartres, the older and, some would say, the more perfect.

When the snows came in winter, the cathedral sat in a valley of whiteness. The massive doors with their brass handles opened outward. In the vestibule there were marble fonts of holy water that rose from the floor. I remember my mother stopping there and crossing herself. I remember the water dripping, splashing against the stone. The great window was a rose of filtered light behind the choir loft. We walked down the aisle of the long nave, genuflected before the main altar, which was set back behind a gold railing, and turned toward the right, into the transept, toward the small altar of the Virgin, with its red vigils, its thin smoke rising into the ribbed vault, and the sweet smell of wax. She was young then, not more than forty. The beads of the rosary were wound loosely about her smooth hands. The bronze crucifix hung from her fingers.

I was a small boy, impatient with sermons, with the words of Ecclesiastes: how the winds turned to the north and then back, how all things returneth to their ordained places. I longed, instead, for the silence that came afterward. For the moment when the congregation had gone into the cold day. When the mass was over, she would light a candle and offer it to the Virgin. You could smell the carbon wax burning; the taper, when it was extinguished, gave off a long streamer into the air. There was barely a sound. Maybe a wooden pew would creak or the sound of the lamp, with its small amplitude, would carry down from the sacristy. A shaft of blue light from the transept windows crept along the altar floor. When it came time to leave—we were always the very last—the streets had filled with snow, knee-deep and powdery and dry, so that it flew up around us and sparkled as we walked. The coalsmoke from the chimneys rose and thinned, disappeared into blue sky.

I remembered that scene as I sat by the river, remembered it as though I could touch it. As though what was here now in the valley, and what was there then, were somehow the same. "It is the same for both of us," I heard Eugene say. "The same mystery all around."

"I understand," I said out loud. But my words were dissolved in the sounds of the river and were quickly carried off to the lake.

The black stone sat before me in the river. By my reckoning it would be there a very long time, so slowly do these silicates dissolve. Down valley, down near Bull Pass, across from where the light met the Matterhorn and the Denton Glaciers, sat the white shells of the sea, among which I had dug only weeks ago. I thought of these two, stone and shell, nothing moving but the river, as though they carried in them, in their deep structures, one of the unlocked secrets of the world.

In our symbolism, as Auden reminds us, the stone has always been reason, so solid and ordered, so self-contained, so muted and strong, so certain of itself. The stone of geometry, of matter linked to form, to the four-cornered tetrahedra of the silicate lattice, the stone, as Wordsworth said, "which holds acquaintance with the stars," with their timeless perfection, their eternal logic. An equation turns the stone.

The shell is different. The shell holds commerce with the sea, comes from the waters of the sea, from its calcium and carbonate. It is the symbol of passion, of the sea's wildness, of its yearning to return, to give back what it possesses, to fling with abandon its treasures untested onto the night shores. It is the forbidding mystery of Curtsinger's dark waters. The shell is inspiration and spontaneity. It is the dream of the sea turned to matter. Wordsworth says of the shell that it has "voices more than all the winds"; that it is also "a joy, a consolation and a hope."

In our literature, the stone has always been mathematics and proof; the shell, poetry and desire. That the two should lie so far apart here in this valley—the dark ventifact on the banks of the Onyx, so near I could take it; the shell distant, at the bend of the river, down by Bull Pass—seemed appropriate, for we have always thought of these as separate and opposed. But it has not always been this way. When I think of the shell and the stone and of all that these convey, I think of a cathedral. And I think of science.

Chartres was begun in the year 1194, about four hundred years after the most recent warming event in Wright Valley. Like all of the French Gothics, it was fashioned from limestone quarried not far from the church itself. Much of the main body of Chartres was completed in the remarkably short period of thirty-one years, and in consequence, it possesses an architectural unity that is rare among European cathedrals.

In the pointed roofs and towers, in the branchlike ribs of the vaults, there is a sense of the whole main structure—nave, transept, and apse—having grown organically out of the earth of the Ile de France. From the croisée, the roof appears to float as though the stone has entirely lost its weight, as though it were dense leafage, "the overhanging boughs of trees," as Eugene once wrote. At first it may seem remarkable that stone could be given such vitality—this massive stone, dragged for miles from the quarries at Berchères-l'Eveque, sometimes by as many as a thousand workers in silent prayer—that it could mediate the space between Earth and sky as though it were united to both. In part this is the work of the light and the stained glass. For from the rose of the west façade and from the three lancets beneath it, there is such a pouring in of light upon the dull limestone walls that it appears like a shifting veil of color, different day to day, hour to hour; blues and reds and greens you cannot imagine. Celestial.

But consider, too, the stone.

Limestone is mostly calcium carbonate, though it is often brushed through with impurities. Pure limestone tends to be white or cream-colored, while those having a high organic content range from gray to black. Limestones that contain finely divided crystalline pyrite are usually tinted gray, and the presence of iron oxides can give rise to muted yellows and reds and browns. Most limestones are an accumulation of marine shells, of mollusks and brachiopods, corals and oysters, the massive skeletons of crinoids, broken and whole, often cemented by the fine calcareous muds of the microscopic forams and algae. The whole undergirding of Ohio, the rocks of the Cincinnatian Series, is nothing but a vast platform of bryozoa and "lamp-shells," horn corals and nautiloids, the stored sediments and memories of the Ordovician sea.

That Chartres should be limestone, that it should be the union of shell and stone and all that those signify, seems fitting. "Whatever else it meant," Henry Adams said, "Chartres expressed an emotion, the deepest man ever felt, the struggle of his own littleness to grasp the infinite." This struggle, this quest to see ourselves in all of our smallness against the timeless sweep of things, to position ourselves in the world, could only be engaged at the highest level, with all that art and science, poetry and technique, passion and logic, working together, could proffer. Not only were the shell and the stone joined as material facts in the limestone of Chartres; they were also joined as the symbols of all that unity could inspire.

Science, too, has its cathedrals, as Gilbert N. Lewis and Merle Randall noted in their preface to the textbook *Thermodynamics*. These are the great theories—the efforts of a few architects and of many laborers. These theories, by which we give order to the world, by which we give it depth and structure, by which we locate and position ourselves against its mystery, are, no less than Chartres, the fusion of inspiration and logic, of discovery and justification, of all that is metaphorically carried by the shell and the stone.

Newton's wild image of a force—the invisible yearning of objects through space, of Earth and Moon, Earth and apple, every hydrogen in the arm of every galaxy tuned to every other—was nothing less than poetry. Such fine and sweeping madness. Such exuberance of vision. Yet the vision had been cast and constrained in logic, in the equation of force and mass and distance. And this equation had to be but a beginning, the generatrix of other equations, other possibilities, of Kepler's laws and Galileo's. And these had to be tested and shown to be right in their domain. The historian I. B. Cohen reminds us that space exploration is but a straightforward application of the gravitational physics of the *Principia*. So we offer these symbols, these concepts, arranged just so, like the mosaic glass of Chartres, and the world responds with its treasures: its ice-mantled moons, its Galilean "stars" so close to the airless ships we can touch them. The brief historic curve of science runs through these points whose coordinates we call dream and vision, logic and proof, whose trace, overall, is the mysterious trace of the world.

When Rutherford imagined the atom as emptiness and speed and force, he imagined it in space, like the architects of Chartres, seeing the figurative towers set apart, seeing the nave like a wide boulevard to God. Bohr's vision was clearer, analogical, precise—the planetlike electrons in the thrall of their nucleus, in their quantized orbits, of just these energies and no others. With Bohr you could see how the cathedral might look, you could imagine its interior, you could imagine how light and matter were conjoined in the space beneath the rose, you could see them related. And then Schroedinger with his orbitals and his rules—the quantum numbers, the clouds of charge. With Schroedinger the cathedral of matter was nearly complete. All that remained were the porches and the statuary, the fine sculpting that would occupy whole generations. But all of those things would take vision on their own scale, and passion and care, and to the stonemasons and artisans those embellishments would seem as important as the cathedral itself.

▼ ▼ ▼

Why, like the makers of Chartres, like the architects and guildsmen of France, like the peasants who dragged the stone in silent prayer, do we build these structures against time? Why do we busy ourselves with these monuments, with these stark homely figures, these robes and faces, these croissures and lambs and garlands? Why, like the farthest-flung mason at the highest tip of the arch, do we care about every line, every roughness? Who, after all, will know? What is it that we build, and why? To what distant choir notes do we listen?

We stand always on the shores of mystery. Once everyone knew this: Newton, Copernicus, Kepler, Galileo, Einstein, Bohr, Schroedinger —every architect of motion and matter. The unknown architects of Chartres knew it too. Beneath the vaulted nave, on the banks of the river, in the stilled transept of remembered youth, among the crowded bays and shelving of the lab it is all the same. We brush against the contours of mystery as though it were a solid thing, as though it were tangible, a curvature, a darkness that is solace and rest, a conduit to what lies beyond. We trace it with our finger the way the glacier traces stone, we take its measure but are lost. What is beyond is all, but what is beyond is hinted at, is eternally present, in what is here—in the swift river and the fierce wind, in the glass, in the melting ice.

It is as though we were destined to wonderment and praise. As though it could be no other way. This endless dragging and piling of stone, this drilling of ice.

The cathedral we now build is so vast and so long in the making that it has had a hundred architects, a hundred thousand workers. Who would name names for fear of leaving out so many? I think of James Hutton and Charles Lyell and Charles Darwin, their sense of time without measure (but it could be measured!), their knowing of the Earth's great patience, of how all things—mountains and deserts and sheets of ice, beings large and small—come and pass away. I think of Rutherford, whose understanding of the very small, of matter's disintegrating core, gave us entry into time, gave us a lamp by which we could light the walls of the foundation as we worked our way downward and back; of Alfred Wegener, who knew—even without knowing how it could ever happen, without knowing what forces might be large enough—that the continents dissembled in their claims of stasis, that, like the Copernican Earth itself, they moved and sundered and built. I think of Wilson and Hess, their vision of Earth and seafloor as sculpture, kinetic on the hot currents of the mantle below, building and destroying, setting islands adrift in the sea; and of Robert Garrels and Rachel Carson, of the connectedness of all things, of light and matter, organic and inorganic, the shell and the stone joined.

This vision of the Earth, this theory by which we gather mountain and sea vent, ice sheet and air, present and past, has been so long in the making. But it is on the verge of becoming a new cathedral. One hears the song of hammers in the transepts, the sound of stone being dragged to the buttresses, the roar of the fire on the lead of the stained glass. One hears winches and pulleys reverberate through the nave high into the stone vaults. There are voices everywhere. And though the construction has far to go, one senses what will be: how the light will fall, how it will break in color on the stones, how the curve of the stone will reach for heaven; how the pilgrims will come to extol.

▼ ▼ ▼

If these journeys and cycles were not science, we would call them myth. If they were structures, unified and whole, rising above the grain of the Ile de France, we would call them Chartres.

Imagine this Earth as though it were a cathedral, only grander, more spacious and lovely, more delicate and complex, yet more robust than we had ever thought, with its thrusts and counterthrusts, its balance and grace. And who are we that we should know these things, that we should gain entry to these holy spaces, dark and lighted, changing with the day? Who are we that we should wonder at all this?

We are the builders, the crafters, the makers of designs. Even now we can hear the hammers singing in the nave. We can see the dust of continents settle. The shape comes clear. The light is abundant and pure. The artisans and stonemasons are at work. The friezes of the lintels take shape. There is so much more to do, so little time, and yet this is a thing of praise.

TWENTY

North and East: Ohio

A man's work is nothing but the slow trek to rediscover through the detours of art those two or three great and simple images in whose presence his heart first opened.
                                                                    Albert Camus

T HE LAKES AND RIVERS WERE CHANGING by the hour. The glass of the moat was slowly closing on itself. The streams were thinning into veins of ice. The winds were picking up, flinging sand across the emptiness of Fryxell. The sun was lower now, riding deeper in the valleys. The crash of summer ice, the sibilance of the streams—these were gone. I wanted to pencil in the coming darkness, the valleys' closing down. Light, temperature, conductivity, dissolved oxygen. But I was tired. My fingers were swollen and yellow. When I touched a knuckle to my tongue, I could taste nitric acid from the pipettes and grains of salt from the lakes. I had lost thirty pounds. Whatever my doubts, it was time to leave, and Dr. Yu and I were on the manifest for the morning flight to Christchurch.

The room was dark and warm when I awoke and climbed out of bed. The blankets had been kicked onto the floor, where they lay in knotted loaves over the flight bags. The map of the continent, with its histories and tracings, still clung to the brown window shade. I walked to it, smoothed it out, ran the palm of my hand over the seas and shelves, over the scallop of the ice, and prepared to fold it and take it home.

On the scale of the map, Victoria Land was a mere stitching, a seam and a border. In our work, we had hardly touched the prairie vastness of the continent beyond the mountains. Yet that vastness had seemed always present to us as power and purity. "A silent and infinite force," Adams had said as he stood before the dynamo in the Great Exposition Hall of 1900. It was like this, something to which you gave obeisance, out of instinct, out of something deep and primal and un-planned. What lay beyond seemed God-struck, the whiteness and blueness of the Earth, its true colors unbrushed, its true sound the sound of wind over ice and water, its true structure the structures of hydrogen and oxygen, linked and bonded and vibrating. In the molec-ular depths of matter the sparkle of ice begins, its layering and layer-ing through time, its freezing and refreezing, its metamorphosis, its astounding hues. And through these, as through the glass of a cathe-dral, come intimations of, and longings for, things as yet unnamed; an emotion of one's smallness, of an overwhelming might.

Still the land evoked more than this. The ice seemed a reminder of the universe at large, of the universe as accident, as matter blown and strewn and expanding, "heartless," as Melville had described it, all moon-filled and dry, hung with poisoned worlds, incinerating stars, vacuums of frozen light. Loneliness, the warm sun as memory, as myth, the blankness of white landscape, in which we see no trace of ourselves, no artifact of our genius and cunning, in which we see none of the greens and oranges and tulip-reds of the "living world," can be terrifying. "The palsied universe lies before us like a leper," Melville also said. And Captain Scott, having finally achieved the object of his heart's desire, the very Pole itself, with its gloried whiteness, its radi-ance in all directions, could say only, "My God, this is an awful place."

I folded the map, placed it in my orange flight bag, and pulled up the shade. The light was dazzling. I had to turn away from it for a sec-ond, avert my eyes. Light from everywhere, from the Ross Sea, water flecked and ice-riven, from the cirques and snowfields across the Sound, from the cone of Erebus reflected a thousand times in the bottle-green bergs and back again. The Society Range seemed balanced against the sea, seemed buttressed and rising, its ribbed ice flung upward in great arches against the sky as though it were holding the whole continent aloft. Through sixty miles of pure air, where stone and glacier and water met, I had never seen such blues and silvers as these. Land of emptiness, land of light.

There were only a few minutes until check-in at Hill Cargo. I was throwing shirts and vests and knotted gloves into bags, emptying drawers of mismatched socks. Varner was laughing at this frenzied attempt at departure. He was joking about how socks never obeyed the law of conservation of matter. "Socks can be created and destroyed. Out of nothing," he said. "So can gloves." He was right. I had two pairs of left-handed mittens and six pairs of unmatched socks. Socks I hadn't even worn.

We had to sit on the flight bags to close them. "Creation from nothing," Varner said, snapping his fingers. "*Ex nihilo.*" And there were still things left in the dresser. Things that had apparently grown there in the dark when I was in the field. I stuffed balaclavas and bear-claw mittens and more socks into my parka and deep into the pockets of my windpants. I was beginning to bulge, to become vegetable-like. "Hey," Varner said. "Check this." And he threw a set of dogtags across the room. I slipped them around my neck.

"When are you going back to Hoare?" I asked.

"The flight's at noon today," he said. "Just me and Walt." He looked eager to get back into the field. Even if it was only to remove the flume.

"I think we did some good things," I said. "Learned a few things no one knew before." He nodded and smiled and said it had been fun. "In a few decades," he said, "I might even understand what it meant for me to be here."

We were at the door. Mike had picked up one of my bags for the walk over to Hill Cargo. "Safe journey," I said.

"I'm not leaving," he responded. Then he handed me an envelope. Inside it I found a sheet of graph paper, with more data plotted on it. "Take a look at this when you're on the plane. You'll like it." I shook his hand, and Mike and I walked down the long hallway to find Dr. Yu.

I glanced at the envelope. On the back he had printed a mailing address:

Larry Varner
Lake Hoare
Antarctica

▼ ▼ ▼

Dr. Yu and I were on the ice runway. The Herc was shuddering. You could smell fuel oil, see its burned vapors disappear as threads of mist along the sea ice. Rimming the horizon, the mountains rolled inland. The frozen Sound, the sky, Erebus on fire, rising on the white shoulder of the world. A small Adelie tobogganed in the distance, slid off into the icy sea.

The Herc is all noise and concussion, throttled energy holding back. The rumble of skis on packed snow, the imperceptible lifting, the last resounding thud of the metal belly against the runway, the long climb, the vibrato of the engines, level flight like a blue flame above the white Earth. Then the inevitable question, posed from a small window overlooking the world: "Will I ever see this again? Will I ever be so blessed?" And to this I have never once been able to answer no, not even the first time. I have always said good-bye in the tentative way of a summer visitor waving from the curb, looking back but looking ahead, too, to the possibility of return. Some places you depart with finality, thankfully, having savored them once; others you can never leave except as an interlude. It is this way.

Pink light bathes the mountains, the flanks of Erebus, the Societies, the newly discovered lands of Ross. The broad ice tongue, like a starched apron laid upon the sea, like a vast wafer, glides off to the south, retreating Poleward, disappearing. A huge berg lies to port. I fix it in my sight, hold it, refuse to let it go. The way you hold on to some-one you love. Its walls are on fire, candlelight in a blue sea. But grad-ually, it seems like hours, it recedes too, off the white stage of the world, and I can no longer look.

Among the hung sleds and survival suits, in the hollowed empti-ness of the Herc returning for cargo, I find a place to lie and stretch out, a place away from the windows where the light is the valley-light of tents. I dream of the ice seas passing beneath me, warming, melting to water, becoming a single phase again, rolling with Earth's curve north toward the islands and home.

▼ ▼ ▼

The entrance to our world is so ornate and rich, who among us would not feel sighted and blessed by it? And yet this was only New Zealand, only the portal. *Have I ever been here before?* I wondered. *Who made all of this?* Everywhere there was the splash of waters, of oars being dipped into the small currents of the Avon. The red kayaks drifted around the bend of the river into the trees. I could hear the sound of fountains. A promenade in the formal garden was floored with butter-cups, lined with sequoias and cypress, the lush branches of the Japanese larch, the plum-colored leaves of the European beech, beneath whose canopy there was darkness in the midst of day. The laughter of children, the sight of women, their movement in dresses along the avenues, came as surprises.

One night it rained. I was under the eaves of the guest house, the window open, the curtains swollen inward from the green park. Drafts of warm summer air, bearing the deep scent of soil and decay and blooming, carried like incense across the fresh bed. And the rain beating as I had not heard it before, turning to rills and rivers on the guttered roof, lying in pools in the sacred dark. And one night there were stars. Dumbstruck, I watched them rise and fall over the garden, the silent seaman's Cross arcing the world, dragging with it the nimbus of galaxies whose names I did not know. The ground was moist and breathing and vivid with smell, every blade and root a pungency mixed with sea air, with the far-flung fields of New Zealand blooming.

Coffee, tea, scones, the smell of sausage in small shops, the sound of swift heels against the pavement, the opening and closing of doors, the crisp unfolding of morning papers, the sound that leaves make, the shadows they cast, minute and detailed, not the shadows of whole mountains, but the shadows of things small and edge-filed—all of this struck me. An engine turned over, purred off into the dawn. A man in a wizard's cap (an alchemist?) wandered through the city square speaking aloud, exclaiming to heaven, his arms thrown wide. Was he mad? Was I? The sun warmed my neck. Warmth from the sky. Was this true? And the winds were only zephyrs here. They spoke among the buildings, through the alleyways, moved bits of paper like spirits before them. I opened my shirt, felt as if I were light enough to float. Before crossing the wide streets, I remembered to look both ways. So much movement, so much speed. All that Varner had dreamed in the valley, and more. The day spoke in a thousand tongues, all at once. There were doorways to enter, lobbies, spaces framed in glass, faces reflected. Everywhere it was green and whispering. I could not see my breath.

We stayed in New Zealand only a few days and then moved east across the date line, the equator, the Pacific. In Honolulu we went to see Keith to tell him what we had learned about the fate of cobalt and lead. "You were right," I said. "The manganese oxides appear to be cleansing the lake."

"So it's all in balance, then," he said. "Input. Output. Steady state?"

"Near as I can tell. Perfect balance. You can just feel the whole thing working out there, turning quiet as a great wheel."

"Cycles," he said. "The whole world moves on them. But I've been thinking more about the atmosphere recently. You know, Charles Keeling set up his station in the early fifties right over there at Mauna Loa." He pointed offhandedly to an invisible space out in the flat Pacific as though he were indicating some feature in his own backyard. "How did your experiments on that little stream down there go? Did they tell you anything about climate? About the atmosphere?"

I scratched my head, laughed a little. Another of Keith's questions. These were always dangerous.

"There's not much of this," he said, lifting his glass against the sweet night air. "This atmosphere is such a tiny fraction of the whole thing. In terms of mass, it is only a few hundredths of a percent of the ocean out there." He motioned toward the Pacific, moved the glass back and forth, creating little whorls that smelled of plumeria. "The atmosphere is nothing. A few deep breaths for the gods. For Kanaka and Pele over there. And yet it's everything."

I was thinking of the graph Varner had given me at McMurdo, thinking of how the stream rose with the day, seemed to awake with it; how it crested and fell, iced over at night. The same pattern over and over, as though the water were trying to tell us something. But what? "Maybe it's like this," Varner had scribbled on the back of the graph. And there were little arrows moving around, connecting "carbon dioxide," "climate," "wind," "dust," "insolation," "albedo," "melting." Who knew what he was getting at? Who knew what the little news clippings they had tacked up at Vanda Station meant? What the increased flow of the Onyx, the swelling of the lakes, meant?

And I was thinking how remarkable it is that what is unseen seems to sing the way the cycles and epicycles of Ptolemy sang; the way the electron clouds of Schroedinger sing. It is remarkable how what is unseen bequeaths a sweet depth and solace upon all things.

From Keith's lanai you could hear this singing, you could sense the whole world working again. It never stopped. This lifting and heaving and release, this purifying that went on and on. The lush forests of the Koolaus behind us inhaling carbon from the thin air, fixing it as rings and chains of glucose, as cellulose and lignin—the rings of Kekule being shaped in secret under the night stars. And the living volcanoes, with their orange lava exhaling far from within the deep Earth. And in the dark sea, the crowded coccoliths and forams bobbing, turning it all silently to stone.

"Well," Keith said, holding his glass in both hands, rotating it slowly so that heat flowed evenly into the ice, transformed it visibly to water, "you've got the first data on that stream, on that glacier. You should have some clear signals in a few years. We all will. We're all so new at this game. It's too big for us right now."

"Too big," Dr. Yu echoed. And we all sat there staring off beyond Diamond Head, beyond the passing cruise ships with their strings of colored lights.

▾ ▾ ▾

That night Dr. Yu and I slept in the small apartment beneath the house. Strong winds blew from the mountains out to sea, threw lariats about the world, yoked volcano to volcano down to the very shores of the Ice. You could hear the palm trees chatter and fall still and then chatter again. Once I thought the roof would be blown away, that I would be lying again under the open sky of the valleys.

In the morning there was a phone call from my sister, Elizabeth. "It's Mother," she said. "She's in the hospital. Come home today. The doctors say there might not be much time."

Keith drove us to the airport. We wove in and out of traffic, past the dead volcanoes with their weathered rims, near the edge of the sea. When we arrived at the gate, I jumped out. I forgot to say good-bye to Keith. I forgot even to look back.

Over the Pacific, I wondered if space could be foreshortened, if time could dilate. *Will I ever see her again?* I asked. There was a whole lifetime in this question. I could not let myself answer no. I was thinking of the Cape, of the shells, of the white mornings that would not come again. Once she had given me a copy of Linus Pauling's great book, *The Nature of the Chemical Bond.* "I heard this is all the rage among you chemists these days," she said. "But you couldn't tell by me." And she laughed. "Maybe you can make sense of it."

I had carried the book with me to college that year, had shown it to Varner. We had both marveled at it, at all the intricate structures it contained: the disk of the carbonate ion, three oxygens splayed out in a plane from the central carbon; the octahedral arrangement of sulfurs in pyrite; the infinite layered structures of the silicates, mica, biotite, kaolinite—Earth stuff. The hydrogen-bonded structures of water and ice. "Pauling, the grand architect," Varner had exclaimed. "How can anyone know all this? I think he actually sees the molecules."

Most of it was beyond me then. But I remember what Pauling had written about water. I read it many times, turned it over in my mind, was utterly taken by the mystery of it: how water retained, like a childhood memory, a trace of its past as ice. How it never forgot that. How it carried that singular fact with it—in its bonds and structures, in its very being—all the way to the boiling point, to where it no longer existed as a liquid. It was water's memory that explained so much, that explained everything, really.

So much went back to that. We were crossing the Grand Canyon. Far below you could see the glowing red of the stone, two billion years of history, lost worlds being slowly recalled by the path of river. I turned away from the window, drew the pecten shell from my pocket. It was rough and pitted. Grains of sand from the Wright Valley still clung to it on skins of moisture. I ran my thumb over it, traced its edges as though I were tracing a continent. Moved it between my thumb and forefinger. A talisman. I imagined giving it to her, imagined I had made it back, saying I found it at the base of the mountains near where the ice had dropped it, near where the ice had carried it in from the sea. It came from water, I said, like the augers and welks, like all those summers past. Calcium and carbonate, sphere and disk, joined.

I was imagining her as she bent over the marble font, blessed herself, the water beading as she made a cross over the heavy coat. Then we went out into the snow, which was dry and blowing and turning the light with its prisms as it fell, no two snowflakes the same. She lifted her face to it. "I think we are younger on these mornings," she said, looking over at me. "Immortal, it feels like. Maybe a little divine." She smiled at this as though it were just a stray thought. Something she did not want to pursue. Something a small boy would not understand anyway. We were near the bottom of the valley the creek had carved. You could see the terraces of its cutting. And against the sky, the gold of the crosses.

I was just thinking this, holding the shell, watching the evening light scatter from it, pure white, except in places a trace of pink, like the snows of Erebus. I was thinking how the world takes and gives back; how it takes again. How things lost become found: the dark hypolimnion in autumn plumes upward like a scroll, bringing to the surface the stored phosphorus and nitrogen, the reduced manganese and iron of summer, the elements of renewal; the flashing coin dropped from the ship carries along the seafloor, is subducted down trenches, heated and released as silver; the calcium and carbonate of the shell return to the quarry, to the cathedral, as stone. I was thinking how water remembers it was ice. On the continent, the ages shook themselves in white, an eternity of strangeness and splendor refracted around you, lying about as ice cap and glacier and spinning berg, as foreign and beautiful as a Jovian moon. And you stood in the thin stone-cut of the mountains, in a place where the world was so new you could breathe its beginnings, the valleys with their lakes and rivers speaking on the wind. And somehow you were back where you had started, back where the first longing, the first recognition began, on those mornings when you had heard it all—in the whisper of snow, in the drop rounded on the fingers and falling, in the smallest bits of matter. When you had heard it all, for the first time, singing.

Eon	Era	Period	Significant Events
			Earth formed around 4.5 bya
Archean 3.8–2.5 bya	Precambrian —to 540 mya		Earliest rocks around 3.8 bya Origin of life (earliest bacteria and algae) around 3.5 bya First stromatolites (particles trapped in microbiotic mats)
Proterozoic 2.5–540 bya			Buildup of free oxygen in atmosphere Eukaryotic cells
Phanerozoic 540 mya-present	Paleozoic ("ancient life") 540–245 mya	Cambrian 540–505 mya	Shelled animals and reef builders Trilobites Earliest chordate (human evolutionary group)
		Ordovician 505–440 mya	Climate becomes more stable Abundant marine algae First vertebrates Limestone deposited near Acton Lake
		Silurian 440–407 mya	Early land plants Early land animals (insects, spiders, scorpions)
		Devonian 407–360 mya	Age of fishes Late Devonian: first four-footed vertebrates
		Carboniferous 360–256 mya	Widespread forests (today's coal deposits) Rise of amphibians First reptiles (by late Carboniferous, mammal-like reptiles dominate the land to end of Triassic) Late Carboniferous: Glaciation in southern hemisphere, including Antarctica
		Permian 256–245 mya	Supercontinent Pangaea is finally assembled First gliding reptiles Antarctic coal deposits around 250 million years ago Mass extinction of marine creatures —nearly 99% of species are wiped out. The end of "Old Life"

Eon	Era	Period	Significant Events
Phanerozoic 540 mya–present (cont.)	Mesozoic ("Age of Reptiles") 245–66 mya	Triassic 245–210 mya	Widespread distribution of *Lystrosaurus* over Gandwanabund, the southern portion of Pangaea Breakup of Pangaea Rise of dinosaurs and first true mammals
		Jurassic 210–144 mya	Dominance of dinosaurs First birds. Breakup of Gandwanaland at end of Jurassic
		Cretaceous 144–66 mya	Continued breakup of superconti-nents and beginning of modern continental distribution First flowering plants in Late Cretaceous.
	Cenozoic ("Age of Mammals") 66 mya–present	Cambrian 540–505 mya	Extinction of dinosaurs at end of Cretacious
		Tertiary 66–1.8 mya	Formation of Antarctic ice cap, causing lower sea levels and modern ocean circulation patterns First upright bipedalism (3.75 million years ago) and first upright hominids (3.6 million years ago)
		Pliocene Epoch 5–2 mya	Ice Ages with subtropical inter-glaciation, including extinctions First members of genus *Homo* (1.6 million years ago) and *Erectus* (1.3 million years ago)
		Quaternary 1.8 mya–present	
		Pleistocene Epoch 2 mya–10,000 years ago	
		Holocene Epoch 10,000 years ago–present	

PERIODIC TABLE OF THE ELEMENTS

1 IA	2 IIA																1 HYDROGEN H 1.01

Metals | Metal-loids | Non-Metals

		3 IIIB	4 IVB	5 VB	6 VIB	7 VIIB	8 VIIIB	9 VIIIB

| 3 LITHIUM Li 6.94 | 4 BERYLLIUM Be 9.01 |
| 11 SODIUM Na 23.0 | 12 MAGNESIUM Mg 24.3 |

19 POTASSIUM K 39.1	20 CALCIUM Ca 40.1	21 SCANDIUM Sc 45.0	22 TITANIUM Ti 47.9	23 VANADIUM V 50.9	24 CHROMIUM Cr 52.0	25 MANGANESE Mn 54.9	26 IRON Fe 55.8	27 COBALT Co 58.9
37 RUBIDIUM Rb 85.5	38 STRONTIUM Sr 87.6	39 YTTRIUM Y 88.9	40 ZIRCONIUM Zr 91.2	41 NIOBIUM Nb 92.9	42 MOLYBDENUM Mo 95.9	43 TECHNETIUM Tc (98)	44 RUTHENIUM Ru 101.1	45 RHODIUM Rh 102.9
55 CESIUM Cs 132.9	56 BARIUM Ba 137.3	57 LANTHANUM La 138.9	72 HAFNIUM Hf 178.5	73 TANTALUM Ta 180.9	74 TUNGSTEN W 183.9	75 RHENIUM Re 186.2	76 OSMIUM Os 190.2	77 IRIDIUM Ir 192.2
87 FRANCIUM Fr (223)	88 RADIUM Ra (226)	89 ACTINIUM Ac (227)	104 RUTHERFORDIUM Rf (257)	105 DUBNIUM Db (260)	106 SEABORGIUM Sg (263)	107 BOHRIUM Bh (262)	108 HASSIUM Hs (265)	109 MRITNERIUM Mt (266)

| 58 CERIUM Ce 140.1 | 59 PRASEODYMIUM Pr 140.9 | 60 NEODYMIUM Nd 144.2 | 61 PROMETHIUM Pm (147) | 62 SAMARIUM Sm 150.4 | 63 EUROPIUM Eu 152.0 |
| 90 THORIUM Th 232.0 | 91 PROTACTINIUM Pa (231) | 92 URANIUM U 238.0 | 93 NEPTUNIUM Np (237) | 94 PLUTONIUM Pu (242) | 95 AMERICIUM Am (243) |

Parentheses denote the best known or most stable isotope of a particular element.

10 VIII B	11 I B	12 II B	13 III A	14 IV A	15 V A	16 VI A	17 VII A	18 Noble Gases
HYDROGEN 1 H 1.01								HELIUM 2 He 4.00
			BORON 5 B 10.8	CARBON 6 C 12.0	NITROGEN 7 N 14.0	OXYGEN 8 O 16.0	FLUORINE 9 F 19.0	NEON 10 Ne 20.2
			ALUMINUM 13 Al 27.0	SILICON 14 Si 28.1	PHOSPHORUS 15 P 31.0	SULFUR 16 S 32.1	CHLORINE 17 Cl 35.5	ARGON 18 Ar 39.9
NICKEL 28 Ni 58.7	COPPER 29 Cu 63.5	ZINC 30 Zn 65.4	GALLIUM 31 Ga 69.7	GERMANIUM 32 Ge 72.6	ARSENIC 33 As 74.9	SELENIUM 34 Se 79.0	BROMINE 35 Br 79.9	KRYPTON 36 Kr 83.8
PALLADIUM 46 Pd 106.4	SILVER 47 Ag 107.9	CADMIUM 48 Cd 112.4	INDIUM 49 In 114.8	TIN 50 Sn 118.7	ANTIMONY 51 Sb 121.8	TELLURIUM 52 Te 127.6	IODINE 53 I 126.9	XENON 54 Xe 131.3
PLATINUM 78 Pt 195.1	GOLD 79 Au 197.0	MERCURY 80 Hg 200.6	THALLIUM 81 Tl 204.4	LEAD 82 Pb 207.2	BISMUTH 83 Bi 209.0	POLONIUM 84 Po (210)	ASTATINE 85 At (210)	RADON 86 Rn (222)

GADOLINIUM 64 Gd 157.3	TERBIUM 65 Tb 158.9	DYSPROSIUM 66 Dy 162.5	HOLMIUM 67 Ho 164.9	ERBIUM 68 Er 167.3	THULIUM 69 Tm 168.9	YTTERBIUM 70 Yb 173.0	LUTETIUM 71 Lu 175.0
CURIUM 96 Cm (247)	BERKELIUM 97 Bk (247)	CALIFORNIUM 98 Cf (249)	EINSTEINIUM 99 Es (254)	FERMIUM 100 Fm (253)	MENDELEVIUM 101 Md (256)	NOBELIUM 102 No (254)	LAWRENCIUM 103 Lr (256)

index

▼

Bellevue Literary Press is devoted to publishing literary fiction and nonfiction at the intersection of the arts and sciences because we believe that science and the humanities are natural companions for understanding the human experience. With each book we publish, our goal is to foster a rich, interdisciplinary dialogue that will forge new tools for thinking and engaging with the world.

To support our press and its mission, and for our full catalogue of published titles, please visit us at blpress.org.

BELLEVUE LITERARY PRESS
New York